Simula SpringerBriefs on Computing

Reports on Computational Physiology

Volume 12

Series Editors

Kimberly J. McCabe, Simula Research Laboratory, Fornebu, Norway

Andrew D. McCulloch, Institute for Engineering in Medicine, University of California San Diego, La Jolla, California, USA

Managing Editor

Rachel Thomas, Simula Research Laboratory, Fornebu, Norway

Editor-in-Chief

Aslak Tveito, Simula Research Laboratory, Fornebu, Norway

In 2016, Springer and Simula launched an Open Access series called the Simula SpringerBriefs on Computing. This series aims to provide concise introductions to the research areas in which Simula specializes: scientific computing, software engineering, communication systems, machine learning and cybersecurity. These books are written for graduate students, researchers, professionals and others who are keenly interested in the science of computing, and each volume presents a compact, state-of-the-art disciplinary overview and raises essential critical questions in the field.

Simula's focus on computational physiology has grown considerably over the last decade, with the development of multi-scale mathematical models of excitable tissues (brain and heart) that are becoming increasingly complex and accurate. This sub-series represents a new branch of the SimulaSpringer Briefs that is specifically focused on computational physiology. Each volume in this series will explore multiple physiological questions and the models developed to address them. Each of the questions will, in turn, be packaged into a short report format that provides a succinct summary of the findings and, whenever possible, the software used will be made publicly available.

More information about this series at https://link.springer.com/bookseries/16669

Kimberly J. McCabe

Editor

Computational Physiology

Simula Summer School 2021 – Student
Reports

Editor
Kimberly J. McCabe
Simula Research Laboratory
Oslo, Norway

ISSN 2512-1677   ISSN 2512-1685 (electronic)
Simula SpringerBriefs on Computing
ISSN 2730-7735   ISSN 2730-7743 (electronic)
Reports on Computational Physiology
ISBN 978-3-031-05163-0   ISBN 978-3-031-05164-7 (eBook)
https://doi.org/10.1007/978-3-031-05164-7

Mathematics Subject Classification (2020): 65M60, 65M06, 92C10, 92C37

© The Editor(s) (if applicable) and The Author(s) 2022. This book is an open access publication.
Open Access This book is licensed under the terms of the Creative Commons Attribution 4.0 International License (http://creativecommons.org/licenses/by/4.0/), which permits use, sharing, adaptation, distribution and reproduction in any medium or format, as long as you give appropriate credit to the original author(s) and the source, provide a link to the Creative Commons license and indicate if changes were made.
The images or other third party material in this book are included in the book's Creative Commons license, unless indicated otherwise in a credit line to the material. If material is not included in the book's Creative Commons license and your intended use is not permitted by statutory regulation or exceeds the permitted use, you will need to obtain permission directly from the copyright holder.
The use of general descriptive names, registered names, trademarks, service marks, etc. in this publication does not imply, even in the absence of a specific statement, that such names are exempt from the relevant protective laws and regulations and therefore free for general use.
The publisher, the authors and the editors are safe to assume that the advice and information in this book are believed to be true and accurate at the date of publication. Neither the publisher nor the authors or the editors give a warranty, expressed or implied, with respect to the material contained herein or for any errors or omissions that may have been made. The publisher remains neutral with regard to jurisdictional claims in published maps and institutional affiliations.

This Springer imprint is published by the registered company Springer Nature Switzerland AG
The registered company address is: Gewerbestrasse 11, 6330 Cham, Switzerland

Preface

Since 2014, we have organized an annual summer school in computational physiology. The school starts in June each year and the graduate students spend two weeks in Oslo learning the principles underlying mathematical models commonly used in studying the heart and the brain. At the end of their stay in Oslo, the students are assigned a research project to work on over the summer. In August the students travel to the University of California, San Diego to present their findings. Each year, we have been duly impressed by the students' progress and we have often seen that the results contain the rudiments of a scientific paper.

Starting in the 2021 edition of the summer school, we have taken the course one step further and aim to conclude every project with a scientific report that passes rigorous peer review as a publication in this new series called *Simula SpringerBriefs on Computing – reports on computational physiology*.

One advantage of this course adjustment is that we have the opportunity to introduce students to scientific writing. To ensure the students get the best introduction in the shortest amount of time, we have commissioned a professional introduction to science writing by Nature. The students participate in a 2-day *Nature Masterclasses* workshop, led by two editors from Nature journals, in order to strengthen skills in high quality scientific writing and publishing. The workshop is tailored to publications in the field of computational physiology and allows students to gather real-time feedback on their reports.

We would like to emphasise that the contributions in this series are brief reports based on the intensive research projects assigned during the summer school. Each report addresses a specific problem of importance in physiology and presents a succinct summary of the findings (8-15 pages). We do not require that results represent new scientific results; rather, they can reproduce or supplement earlier computational studies or experimental findings. The physiological question under consideration should be clearly formulated, the mathematical models should be defined in a manner readable by others at the same level of expertise, and the software used should, if possible, be made publicly available. All reports in this series are subjected to peer-review by the other students and supervisors in the program.

We would like to express our gratitude for the very fruitful collaboration with Springer -Nature and in particular with Dr. Martin Peters, the Executive Editor for Mathematics, Computational Science and Engineering.

The editors of *Simula SpringerBriefs on Computing – reports on computational physiology*:

Oslo, Norway   *Kimberly J McCabe*
March 2022   *Rachel Thomas*
                                                                                              Andrew D McCulloch
                                                                                                  Aslak Tveito

Acknowledgements

The Simula Summer School in Computational Physiology is a team effort, with many scientists contributing their time to give lectures and advise projects for the students. The 2021 school was particularly challenging, as it was conducted entirely online and required creative problem solving to allow for effective project collaboration, student engagement, and knowledge transfer. We would like to thank the lecturers and project advisors for their expertise and willing participation in the course: Dr. Hermenegild Arevalo, Dr. Gabriel Balaban, Dr. Jonas van den Brink, Marius Causemann, Dr. Andrew Edwards, Dr. Nickolas Forsch, Dr. Ingeborg Gjerde, Dr. Glenn Lines, Dr. Molly Maleckar, Dr. Kimberly McCabe, Dr. Andrew McCulloch, Denis Reis de Assis, Dr. Joakim Sundnes, Abigail Teitgen, Dr. Aslak Tveito, Dr. Daniela Valdez-Jasso, Jonas Verhellen, and Dr. Samuel Wall. Administrative support for the school was provided by Dr. Kimberly McCabe, Dr. Rachel Thomas, Elisabeth Andersen, and Hanie Tampus.

We would also like to acknowledge the teaching assistants for the course, who provided much-needed support during virtual lectures: Åshild Telle, Eina Jørgensen, and Hanna Borgli. We received training and equipment relating to virtual learning from Simula Kodeskolen, and would specifically like to acknowledge Matteus Häger, Håkon Kvale Stensland, Elin Backe Christophersen, and Marianne Aasen.

The Simula Summer School in Computational Physiology is supported through the Simula-UiO-UCSD Research PhD Training Programme (SUURPh), an endeavour funded by the Norwegian Ministry of Education and Research. Additional financial support is derived from SIMENT, an INTPART mobility grant from the Norwegian Research Council. The school also received funding from Digital Life Norway (DLN).

Contents

7   Investigating the Multiscale Impact of Deoxyadenosine Triphosphate (dATP) on Pulmonary Arterial Hypertension (PAH) Induced Heart Failure   77

Kristen Garcia, Marcus Hock, Vikrant Jaltare, Can Uysalel, Kimberly J McCabe, Abigail Teitgen, Daniela Valdez-Jasso

8   Identifying Ionic Channel Block in a Virtual Cardiomyocyte Population Using Machine Learning Classifiers   91

Bjørn-Jostein Singstad, Bendik Steinsvåg Dalen, Sandhya Sihra, Nickolas Forsch, Samuel Wall

Chapter 1
A Pipeline for Automated Coordinate Assignment in Anatomically Accurate Biventricular Models

Lisa Pankewitz[1], Laryssa Abdala[2], Aadarsh Bussooa[1], Hermenegild Arevalo[1]

1 – Simula Research Laboratory, Norway
2 – University of North Carolina, USA

Abstract There is an increased interest, in the field of cardiac modeling, for an improved coordinate system that can consistently describe local position within a heart geometry across various distinct geometries. A newly designed coordinate system, Cobiveco, meets these requirements. However, it assumes the use of biventricular models with a flat base, ignoring important cardiac structures. Therefore, we extended the scope of this state-of-the-art biventricular coordinate system to work with various heart geometries which include basal cardiac structures that were previously unaccounted for in Cobiveco. First, we implemented a semi-automated input surface assignment for increased accessibility and reproducibility of assigned coordinates. Then, we extended the coordinate system to handle more anatomically accurate biventricular models including the valve planes, which are of great interest when modeling diseases that manifest themselves in the basal area. Furthermore, we added the functionality of mapping vector data, such as myocardial fiber orientations, which are crucial for replicating the anisotropic electrical propagation in cardiac tissue.

1.1 Introduction

The representation of cardiac geometry independent of patient origin and the flawless transfer between different measuring modalities are important tools in clinical research [1, 2]. To accurately describe a local position within the heart, a robust coordinate system is required. Such a coordinate system enables a variety of applications, including the transfer of data between different heart geometries and comparing data produced using different measuring modalities, such as validating simulations with clinical data [2, 3].

© The Author(s) 2022
K. J. McCabe (ed.), *Computational Physiology*, Simula SpringerBriefs
on Computing 12, https://doi.org/10.1007/978-3-031-05164-7_1

A recently published biventricular coordinate system, Cobiveco, offers a consistent and reliable approach for describing positions in biventricular heart models [2]. However, the current state-of-the-art coordinate system is limited to biventricular heart geometries, which are clipped at a specific planar position, such that the resulting base appears completely flat. Clipping the base in this manner also clips the underlying ventricles. Although this clipping procedure remains part of the common mesh generation approach, it does not yield anatomically accurate cardiac meshes for the purpose of computer simulations. We argue that biophysical simulations of the heart should include the clipped base cardiac structures, that contain the valve openings, for more realistic results. The ventricles, together with the presence of valve planes, are important features of ventricular anatomy that can influence cardiac electrophysiology and mechanics.

Features that are connected to the valves are critical anatomical structures, such as the papillary muscle and chordae. Any structural defects that change the shape of the ventricles and alter the activation in the aortic valve annulus can have an effect on electrical dyssynchrony or ventricular dilation. Therefore, the inclusion of valve planes in cardiac models is necessary. This is especially true when modeling certain disease phenotypes, where changes in anatomy, mechanics, and activation manifest themselves in areas closer to the valve planes. An important example of this is congenital heart defect (CHD), which is the most common birth defect worldwide [4, 5]. Heterogeneous morphology and physiology in CHD patients have been shown to complicate risk assessment of individual patients requiring anatomically accurate models. This is a use-case where the inclusion of valve planes in the biventricular models may lead to enormous improvement of the model quality as morphological changes as well as scar tissues in this patient group can be located close to the base.

1.2 Methods

In this work, we extend the open-source MATLAB implementation of Cobiveco for tetrahedral meshes to take into account anatomically more accurate biventricular meshes that include valve planes instead of a flat, clipped base. First, we provide a surface extraction tool that automatically creates input surfaces files required for setting up the biventricular coordinate system. Then, we adapt the existing Cobiveco framework to allow for more anatomically correct geometries. Last, we extend the software to allow for mapping and transfer of vector data between different heart geometries.

1.2.1 Semi-Automated Surface Extraction

Existing tools extract surfaces from meshes and imaging data. Image-based surface extraction operates directly on raw clinical imaging to identify cardiac structures, ventricles follow the same parametrization. This is also reflected in the shared apex definition. To construct the coordinate system, only landmarks which are consistent throughout variations in different geometries, are chosen.

1.2.2.1 Creation of the Coordinate System Cobiveco

A detailed description of the steps involved in the creation of the original Cobiveco framework can be found in [2]. In short, the construction of the coordinate system includes eight steps and is summarized below.

As with Cobiveco 1.0, Cobiveco 2.0 requires a biventricular volume that includes a base containing the four heart valve annuli, including the connecting bridges. Besides the volume mesh, five boundary surfaces as shown in Figure 1.3 are required as input, which is one additional surface compared to Cobiveco 1.0. The surfaces required are a *basal surface* S_{Base}, a *basal epicardial surface* $S_{Epi,base}$, an *epicardial, non basal surface* $S_{Epi, nonbase}$, an *LV endocardial surface* S_{LV} and an *RV endocardial surface* S_{RV}. The utilities for the semi-automated input-file generation are described in Section 1.2.1.

Transventricular Coordinate (tv)

The transventricular coordinate is calculated as described in the original publication [2].

Extraction of Septal Surface and Curve

The *septal surface* S_{Sept} and the *septal curve* C_{Sept} are extracted as described in [2].

Transmural Coordinate (tm)

The calculation of the transmural coordinate follows the same steps as in Cobiveco 1.0, but takes into account the two epicardial surfaces. As we split the epicardial surface into a non-base epicardial surface and a basal epicardial surface, the whole epicardial surface is defined be the union of both, as given in (1.2):

$$S_{Epi} = S_{Epi\_non\_base} \cup S_{Epi\_base} \tag{1.2}$$

Heart Axes and Apex Point

The definition of the heart axes and apex point mainly follows the steps described in the original publication [2]. As the definition of the orthogonal heart axes largely

depends on the truncated septal surface, the calculation of the truncation needed to
be revised to take into consideration the increased curvature of the septal surface
at the base of the anatomically accurate biventricular heart model. Therefore the
septal surface is additionally truncated by 15% at the basal side, where the distance
is based in the direction of $\mathbf{v_{LongAx}}$. This modification results in the final truncated
septal surface $S_{SeptTrunc}$ being calculated by (1.3), where P_q refers to the qth percentile.

$$S_{\text{SeptTrunc}} = \left\{\mathbf{x} \in S_{\text{Sept}} \mid \mathbf{x} \cdot \mathbf{v}_{\text{AP}} > P_{20}\left(\mathbf{x} \cdot \mathbf{v}_{\text{AP}}\right) \text{ and} \right. \tag{1.3}$$

$$\mathbf{x} \cdot \mathbf{v}_{\text{AP}} < P_{90}\left(\mathbf{x} \cdot \mathbf{v}_{\text{AP}}\right) \text{ and}$$

$$\left. \mathbf{x} \cdot \mathbf{v}_{\text{long}} > P_{15}\left(\mathbf{x} \cdot \mathbf{v}_{\text{long}}\right)\right\}$$

Since the *septal curve* needs to be split in two segments, the new geometry with a
closed base requires a different solution than in Cobiveco 1.0 as well. Hence, we
exclude the new, *basal epicardial surface* from the epicardial definition to allow for
a separation of the anterior and posterior part of the septal curve.

$$C_{\text{Sept}} = \left\{\mathbf{x} \in S_{\text{Epi,base}} \mid u_v(\mathbf{x}) = 0.5\right\} \tag{1.4}$$

Extraction of Ridge Surfaces

As the more anatomically accurate biventricular models contain a closed surface at
the base, we needed to modify the ridge definition in Cobiveco 2.0. In Cobiveco
2.0 we aim to replicate the original ridge assignment but use the valve planes
as guiding points resulting in a symmetric ridge. The ridge is used to provide a
boundary condition for the rotational coordinate. Currently, the ridge is defined
from the posterior interventricular junctions, where both ventricles symmetrically
impose a boundary condition, via the mitral valve and tricuspid valve in the LV/RV
septum respectively. The anterior ridge definition is defined via the mitral valve and
pulmonary valve, resulting in the ridge definition as shown in Figure 1.2.

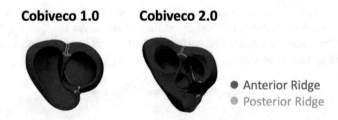

Fig. 1.2: Ridge definition in Cobiveco 1.0 and Cobiveco 2.0. The anterior part of the
ridge is colored in red, while the posterior part of the ridge is highlighted in grey.

Therefore, the ridge is obtained by defining the solution to Laplace's equation with boundaries applied to be 0 on the epicardial basal surface and 1 on the septal surface. The *non-base epicardial surface* is now excluded as a boundary condition, as shown in (1.5).

$$\Delta u_{\text{Ridge}}(V) = 0 \quad \text{with} \quad u_{\text{Ridge}}\left(S_{\text{Epi}} \backslash S_{\text{Sept}}\right) = 0 \quad \text{and} \quad u_{\text{Ridge}}\left(S|_{\text{Sept}}\right) = 1$$
$$(1.5)$$

The solution to Laplace's equation is calculated as shown here in (1.6).

$$\Delta u_{\text{Ridge}}(V) = 0 \quad \text{with} \quad u_{\text{Ridge}}\left(S_{\text{Epi,nonbase}} \backslash S_{\text{Sept}}\right) = 0 \quad \text{and} \quad u_{\text{Ridge}}\left(S|_{\text{Sept}}\right) = 1$$
$$(1.6)$$

The second step remains as set up in the original Cobiveco 1.0. However, the resulting ridge cannot be applied as it is. Currently, a manual filtering step is involved as there remains a ridge within the RV in-between the tricuspid and the pulmonary valve, which is not used as a boundary condition when setting up the rotational coordinate.

The Rotational Coordinate (r)

The rotational coordinate is defined as described in [2].

Computation of the Apicobasal Coordinate (ab)

The apicobasal coordinate is calculated as described in the original Cobiveco article. Currently, however, we used the solution to Laplace's equation as a place holder, as the rotational coordinate still includes discontinuities which prohibit the assignment of the apicobasal coordinate as described in Cobiveco 1.0

1.2.3 Mapping Vector Fields

The original Cobiveco implementation has a scalar field mapping functionality available. To map a scalar field from the source mesh B to a target mesh A, it constructs a matrix $M_{A \leftarrow B}$ from the nodes of the source mesh to the nodes of the target mesh [2]. The user can choose between linear and nearest-neighbor interpolation.

Mapping vector fields is of interest since data, such as muscle fiber fields, are crucial to advance the cardiovascular computational simulations field. Here we enable the functionality of mapping such fields by treating each coordinate as a scalar field. More specifically, the vector field is represented as a matrix of the nodes of the source mesh by three. Each of its columns represents the coordinate of the vector field in each source node. The end result of the vector mapping process is shown in Figure 1.5.

1.3 Results

In this project, we have successfully founded the basis for extending Cobiveco to include more anatomically accurate biventricular models. The results are presented in three steps, namely a pre-processing, processing, and post-processing step. Each step reduces the manual manipulation of meshes, generalizes the biventricular coordinates, and enables vector data transfer, respectively.

First, we successfully implemented a semi-automated surface extraction method that uses a minimal set of parameters, based on a seed point and angular threshold, to identify structures of interest in cardiac meshes. The resulting surfaces, after extraction, are shown in Figure 1.3.

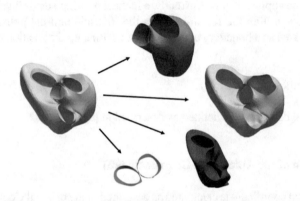

Fig. 1.3: Extracted surfaces after using BFS algorithm and angular threshold.

Second, we adapted the previous Cobiveco framework to work with biventricular geometries, which includes the four cardiac valve planes. The preliminary results of Cobiveco 2.0 are shown in Figure 1.4. The rotational coordinate suffers from inconsistencies at the septum, owing to the manual exclusion of the ridge boundary. Currently, the apicobasal coordinate is only represented by the solution to Laplace's equation. Last, we adapted the framework to include the mapping of vector data. The result for mapping synthetic data is shown in Figure 1.5.

1.4 Conclusion

In this project, we present an updated version of the consistent biventricular coordinates introduced by [2]. The pipeline can be applied to biventricular geometries for mapping scalar and vector data between different hearts.

Cobiveco 2.0 builds upon the original Cobiveco [2], by extending the coordinates for biventricular geometries that include the ventricular base. We aim to keep the

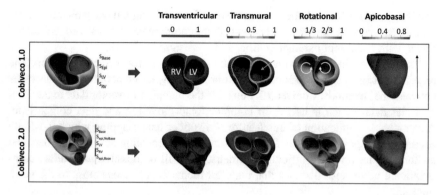

Fig. 1.4: Visual Comparison of the four coordinates created by Cobiveco 2.0 and Cobiveco 1.0. The apicobasal coordinate shown for Cobiveco 2.0 is represented by the solution to Laplace's equation.

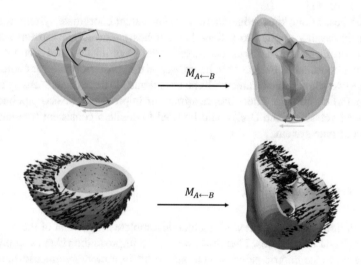

Fig. 1.5: Cobiveco 2.0 has the functionality of mapping vector fields. First, the coordinates are built in the source B and target A biventricular geometries (top). Then the map $M_{A \leftarrow B}$ is used to map vector fields (bottom).

resulting coordinates with the same properties from the original ones: bijective, continuous (apart from the binary transventricular coordinates), normalized, complete, linear, with consistent parametrization, and consistent landmarks.

The new pipeline reduces manual mesh manipulations for surface extraction. The required surfaces can be effortlessly extracted using fewer parameters than other conventional methods. Cobiveco 2.0 enables the mapping of vector data in addition to scalar data, which is useful for computational modeling and data comparison. A remarkable application of this feature is myofiber data mapping which is widely used in electrophysiology simulations. Moreover, by enabling data transfer between anatomical accurate biventricular geometries, it will be possible to validate computational models for diseases that manifest themselves in areas close to the valve planes.

The newly developed pipeline, with the inclusion of the valve planes in the cardiac model, is of special interest for studying the most common form of CHD, namely Tetralogy of Fallot (ToF). In ToF patients, the scar tissue is located close to the base, rendering electrophysiological simulations feasible with our pipeline [6, 7, 8, 9, 10, 11].

To the best of our knowledge, there is no consistent coordinate system available that can be readily applied to a four-chamber heart model. The current state-of-the-art coordinates for the atria, Universal Atrial Coordinates (UAC) is not based on circular coordinates but rather in lateral-septal and posterior-anterior coordinates [12]. The two-dimensional framework can be extended to three dimensions by adding a transmural coordinate. Therefore, merging our improved Cobiveco pipeline with an updated version of the UAC could be used to create a consistent four-chamber heart coordinate system.

1.4.1 Limitations

The limitations of the current work include incomplete assignment of the rotational and apicobasal coordinates. Therefore, we aim to improve the ridge assignment to obtain a more symmetric ridge, which will result in a more symmetric rotational coordinate, as the ridge defines the boundaries set in the rotational coordinate. To achieve this goal, we will modify the anterior part of the ridge in the LV to be defined via the aortic valve and not via the mitral valve as the definition is set now. This will ensure a symmetric set up of the rotational coordinate in the RV and LV. The remaining parts of the proposed framework, however, were successfully tested in one patient-specific geometry.

Besides, a statistical analysis of the errors using a cohort of geometries is necessary to ensure this is a reliable coordinate system. Moreover, the mappings were performed using artificial data. Future work could include transferring experimental data between two geometries to ensure the results are physiologically consistent. Furthermore, the post-processing could also feature tensor data mapping.

References

1. Matthijs Cluitmans, Dana H Brooks, Rob MacLeod, Olaf Dössel, María S Guillem, Peter M van Dam, Jana Svehlikova, Bin He, John Sapp, Linwei Wang, et al. Validation and opportunities of electrocardiographic imaging: from technical achievements to clinical applications. *Frontiers in physiology*, 9:1305, 2018.
2. Steffen Schuler, Nicolas Pilia, Danila Potyagaylo, and Axel Loewe. Cobiveco: Consistent biventricular coordinates for precise and intuitive description of position in the heart–with matlab implementation. *arXiv preprint arXiv:2102.02898*, 2021.
3. Jason Bayer, Anton J Prassl, Ali Pashaei, Juan F Gomez, Antonio Frontera, Aurel Neic, Gernot Plank, and Edward J Vigmond. Universal ventricular coordinates: A generic framework for describing position within the heart and transferring data. *Medical Image Analysis*, 45:83–93, 2018.
4. Bryan Boling. *Cardiothoracic Surgical Critical Care, An Issue of Critical Care Nursing Clinics of North America, E-Book*, volume 31. Elsevier Health Sciences, 2019.
5. Christian Apitz, Gary D Webb, and Andrew N Redington. Tetralogy of fallot. *The Lancet*, 374(9699):1462–1471, 2009.
6. R Bedair and X Iriart. Educational series in congenital heart disease: Tetralogy of fallot: diagnosis to long-term follow-up. *Echo Research and Practice*, 6(1):R9–R23, 2019.
7. Sara Piran, Anne S Bassett, Jasmine Grewal, Jodi-Ann Swaby, Chantal Morel, Erwin N Oechslin, Andrew N Redington, Peter P Liu, and Candice K Silversides. Patterns of cardiac and extracardiac anomalies in adults with tetralogy of fallot. *American heart journal*, 161(1):131–137, 2011.
8. Sotiria C Apostolopoulou, Athanassios Manginas, Nikolaos L Kelekis, and Michel Noutsias. Cardiovascular imaging approach in pre and postoperative tetralogy of fallot. *BMC cardiovascular disorders*, 19(1):1–12, 2019.
9. Ariane J Marelli, Andrew S Mackie, Raluca Ionescu-Ittu, Elham Rahme, and Louise Pilote. Congenital heart disease in the general population: changing prevalence and age distribution. *Circulation*, 115(2):163–172, 2007.
10. Wei Hui, Cameron Slorach, Andreea Dragulescu, Luc Mertens, Bart Bijnens, and Mark K Friedberg. Mechanisms of right ventricular electromechanical dyssynchrony and mechanical inefficiency in children after repair of tetralogy of fallot. *Circulation: Cardiovascular Imaging*, 7(4):610–618, 2014.
11. Charlotte Brouwer, Gijsbert FL Kapel, Monique RM Jongbloed, Martin J Schalij, Marta de Riva Silva, and Katja Zeppenfeld. Noninvasive identification of ventricular tachycardia–related anatomical isthmuses in repaired tetralogy of fallot: What is the role of the 12-lead ventricular tachycardia electrocardiogram. *JACC: Clinical Electrophysiology*, 4(10):1308–1318, 2018.
12. Caroline H Roney, Ali Pashaei, Marianna Meo, Rémi Dubois, Patrick M Boyle, Natalia A Trayanova, Hubert Cochet, Steven A Niederer, and Edward J Vigmond. Universal atrial coordinates applied to visualisation, registration and construction of patient specific meshes. *Medical image analysis*, 55:65–75, 2019.

Open Access This chapter is licensed under the terms of the Creative Commons Attribution 4.0 International License (http://creativecommons.org/licenses/by/4.0/), which permits use, sharing, adaptation, distribution and reproduction in any medium or format, as long as you give appropriate credit to the original author(s) and the source, provide a link to the Creative Commons license and indicate if changes were made.

The images or other third party material in this chapter are included in the chapter's Creative Commons license, unless indicated otherwise in a credit line to the material. If material is not included in the chapter's Creative Commons license and your intended use is not permitted by statutory regulation or exceeds the permitted use, you will need to obtain permission directly from the copyright holder.

Chapter 2
3D Simulations of Fetal and Maternal Ventricular Excitation for Investigating the Abdominal ECG

Julie Johanne Uv[1], Lena Myklebust[1], Hamid Khoshfekr Rudsari[2,3], Hannes Welle[4], Hermenegild Arevalo[1]

1 – Simula Research Laboratory, Norway
2 – University of Oslo, Norway
3 – Oslo University Hospital, Norway
4 – Karlsruhe Institute of Technology, Germany

Abstract Congenital heart disease (CHD) is a leading cause of infant death. To diagnose CHD, recordings from abdominal fetal electrocardiograms (fECG) can be used as a non-invasive tool. However, it is challenging to extract the fetal signal from fECG recordings partly due to the lack of data combining fECG recordings with a ground truth for the fetal signal, which can be obtained by using a scalp electrode during delivery. In this study, we present a computational model of a pregnant female torso, in which we simulate fetal and maternal ventricular excitation during sinus rhythm to derive fECGs, so as to enable isolated measurement of the fetal and maternal signal contributions. To extract the fetal contribution from a combined signal, we apply an adaptive filtering algorithm to wavelet transformed signals. Further development of the model may enable improvements in the recording and processing capabilities for fECGs, the reliable estimation of fetal heart rates, and possibly interpretation of fetal signal morphologies that could improve the overall diagnostic significance of abdominal fECGs.

2.1 Introduction

Congenital heart disease (CHD) is a common birth defect referring to abnormal function or structure of the heart [1]. It arises in early pregnancy stages during heart development [2]. With a prevalence of about 7–10 per 1000 live births, CHD is a leading cause of infant death [1, 3]. For children who survive with CHD, the odds of developing mental disabilities is 9 times higher than in the general population,

© The Author(s) 2022
K. J. McCabe (ed.), *Computational Physiology*, Simula SpringerBriefs on Computing 12, https://doi.org/10.1007/978-3-031-05164-7_2

while the odds of experiencing health problems that limit physical activity is 14 times higher [4].

CHD may be diagnosed during pregnancy, which has been found to decrease mortality risk and improve patient outcome following post natal surgery [5]. The most common fetal monitoring technique today is Doppler ultrasound, either through a single hand-held Doppler ultrasound transducer or cardiotocography (CTG). CTG and hand-held Doppler ultrasound provide estimates of the fetal heart rate (FHR), but no information on the fetal heart rate variability (FHRV) or more sophisticated analysis of the electrical cardiac activity [6]. In addition, CTG has technical limitations such as low sampling frequency, which limits its FHR estimation capability [7]. Since the first commercial fetal heart monitor was launched in 1968 [8], continuous CTG monitoring during delivery has been linked to an increase in caesarean sections and instrumental births, which is a risk to mother and fetus [9], but did not have an impact on perinatal mortality and morbidity [10].

The waveform of the electrocardiogram (ECG) may contain important information about the cardiac electric activity beyond the FHR [11], which is routinely interpreted to assess cardiac function in adults [12]. Several studies have pointed out the need of ECG waveform analysis when monitoring pregnancies [7, 13]. In particular, ECG waveform analysis may be used in addition or as an alternative to CTG [14], to improve the detection of pathologic cardiac events or to provide information on uterine contraction [13].

Despite the above-mentioned advantages, abdominal ECGs show a low signal to noise ratio due to the small myocardial volume of the fetus and interference from e.g. the maternal heart, motion artifacts and muscle contractions. Therefore, postprocessing is required to extract the fetal signal contribution. This is challenging, because signal and artifacts are overlapping in the frequency domain [15, 16]. State of the art methods use adaptive filtering of the abdominal signal with a thoracic reference signal to overcome this problem [17, 18]. A lack of limited gold standard data, comprising simultaneous fECG recordings from abdominal electrodes, and an invasive scalp electrode further complicate the extraction [19]. Previous work has addressed this challenge by creating a fECG simulator in which each heart is represented by a moving dipole [20]. Herein, we simulated fetal and maternal cardiac activity using an image-based finite element model to derive a realistic abdominal fECG.

2.2 Methods

2.2.1 Geometrical mesh construction

For our maternal and fetal heart we used a biventricular mesh of a female adult heart based on CT images [21]. We used myocardial fiber orientations from the same study, which have been generated using a rule-based approach to reproduce experimental findings [22, 23]. Both hearts were then augmented to fit into a female pregnant torso from the FEMONUM database [24, 25, 26]. For the maternal heart, the biventricular triangular surface mesh was translated, scaled and rotated by matching the surrounding torso to the pregnant torso which it was incorporated into. For the fetal heart, the same was done for a fetal torso originally included in the model from the FEMONUM database. Incorporation of biventricular surface meshes into the pregnant torso was done in Paraview [27], while volume mesh generation was done using gmsh [28]. The final model is a 3D finite element mesh with ~ 17 million nodes as shown in Fig. 2.1a. Elements are tetrahedral with sizes approximately 0.4 and 50 mm for the hearts and torso respectively.

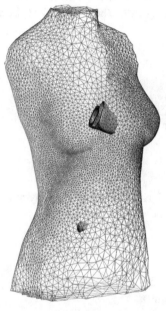

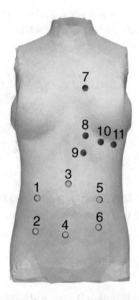

(a) Visualisation of the finite element mesh used in our simulations. Hearts and torso are based on models by [21] and the FEMONUM database [24, 25, 26] respectively.

(b) The extracellular potential was measured at 11 nodes on the torso to obtain fECG traces. The figure shows thoracic (purple) and abdominal (green) electrodes. Numbers denote the different channels used for signal processing.

Fig. 2.1

2.2.2 Electrophysiological modelling

Ionic current properties of the maternal heart were modelled using the Ten Tuss-cher model of human ventricular myocytes [29]. For the fetal heart a modified version of the model, adapted to match fetal ventricular electrophysiology, was used [30]. Tissue propagation was modelled with the pseudo-bidomain approach [31] as implemented in CARPentry [32]. Following [22], intracellular and extracellular conductivities in the myocardium were set to $g_i = (0.27, 0.081, 0.045)$ and $g_e = (0.9828, 0.3654, 0.3654)$ S m^{-1} in the longitudinal, transverse and normal direction respectively, while the torso was assigned an isotropic conductivity of 0.216 S m^{-1}.

Sinus rhythm was simulated over 5 seconds by pacing both hearts with a 10 ms stimulus of 100 uA/cm^2. The stimuli were delivered to all vertices on the endocardial surfaces through a transmembrane current, resulting in the activation pattern displayed in Fig. 2.2. Basic cycle lengths were set to 500 and 450 ms for the maternal and fetal heart, respectively. The fetal cycle length was chosen to reflect the heart rate after 38 weeks of gestation [30].

Fig. 2.2: Activation map of the maternal heart during simulated sinus rhythm. Activation times are measured from stimuli onset and encoded by colour.

2.2.3 Extracellular potential measurements

In order to compute extracellular potential traces using CARPentry, the average potential was set to zero. For the fECG analysis, we located 11 virtual electrodes on the torso, considering both a 9-electrode and 8-electrode fECG setup [33]. The placement of the virtual electrodes is shown in Fig. 2.1b. Note the distinguished abdominal and thoracic electrodes which are meant to record the fECG and maternal ECG, respectively. The extracellular potential for the selected nodes was extracted. Subsequently, the voltage between all electrodes and a reference electrode on the lower abdomen was calculated from their potential difference to obtain ECG signals.

2.2.4 Fetal ECG extraction using signal processing methods

In this part, we study the signal processing methods for extracting the fECG from maternal abdominal recordings. An fECG is mixed with several disturbance sources in the maternal abdominal ECG. The primary source of disturbance is the maternal ECG which coincides with fECG in time, and frequency domains [16]. Hence, the extraction of fECG is challenging, and novel signal processing methods are demanded for the efficient diagnosis of fetal cardiac disorders and further treatment.

Different signal processing methods for fECG extraction are given in the literature. Rasti-Meymandi *et al.* proposed a deep learning algorithm named AECG-DecompNet which efficiently extracts both fECG and maternal ECG [34]. AECG-DecompNet has two main network series; one network estimates the maternal ECG, and the other network eliminates the noise and interference. The two series of networks have shown superior results in QRS (a complex in the ECG signal) detection compared to utilizing one network only. Nevertheless, proper training of the first network is important for robust fECG extraction. Independent component analysis (ICA) [35] and multi-ICA [36] also have been considered for fECG extraction, for which statistical models in the algorithms are challenging. Furthermore, adaptive filtering has also been used for extraction of fECG due to fast and simple implementation [37, 38, 17, 18]. However, fECGs extracted by adaptive filtering still have remaining maternal signal components.

We use an algorithm based on the combination of wavelet analysis and adaptive filtering proposed in [39]. We first process abdominal and thoracic signals by wavelet transformation. Then, we process the detail coefficients of the wavelet transformed signals as inputs for adaptive filtering using the least mean square (LMS) filtering. We subsequently denoise the output of the adaptive filter, and finally, take the inverse wavelet transform of the denoised coefficients.

The abdominal ECG is a non-stationary signal. We use stationary wavelet transform (SWT) to avoid the Gibbs phenomenon by removing down-sampling and up-sampling coefficients. The output of the SWT has the same length as the input signal. SWT has two sets of functions: the scaling and wavelet functions, denoted by $\phi(n)$ and $\Phi(n)$, where n stands for the nth sample point of the signal. The functions are defined based on a chosen wavelet function. We decompose the signal $f(n)$ using the wavelet decomposition to approximation coefficient $c_{j,k}$ and detail coefficient $d_{j,k}$. The coefficients at the jth scale are [39]

$$c_{j,k} = < f(n), \Phi_{j,k}(n) >, \qquad (2.1)$$

$$d_{j,k} = < f(n), 2^{-\frac{j}{2}} \phi_{j,k}(2^j n - k) >, \qquad (2.2)$$

with

$$< f(n), \Phi_{j,k}(n) > = \int_{-\infty}^{\infty} 2^{-\frac{j}{2}} f(n) \Phi \star (2^j n - k) \mathrm{d}n, \qquad (2.3)$$

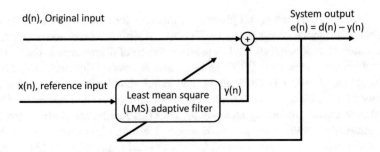

Fig. 2.3: Adaptive filtering for fECG extraction.

where $\star$ denotes the complex conjugation.

At the next step, we consider the detail coefficients of the SWT for adaptive filtering. Fig. 2.3 shows the adaptive filtering used in this section. We utilize the coefficients of the abdominal ECG as the original input $d(n)$ and the coefficients of thoracic ECG as the reference input $x(n)$. The adaptive filter is based on LMS, which its update is given by [39]

$$w(n+1) = w(n) + 2\mu e(n)x(n), \qquad (2.4)$$

where $w(n)$ and μ are the adaptive filter coefficients and step size. Also, $e(n)$ is the feedback in the adaptive filter where the coefficients of $w(n)$ are constantly adjusted until the output $y(n)$ is very close to the maternal ECG components of the abdominal ECG. The output signal $y(n)$ is filtered as

$$y(n) = \sum_{i=0}^{m-1} w_i x(n-i) = w^T(n)x(n), \qquad (2.5)$$

where $w^T(n)$ denotes the transposed vector of $w(n)$. Finally, we consider the output $e(n)$ for denoising at the next step. After denoising using wavelet transform, we take an inverse SWT to obtain the fECG.

2.3 Results

The conducted simulations resulted in distributions of the body surface potential as depicted in Fig. 2.4. When stimulated, both hearts produced regions on the torso of increased potential in the basal direction and decreased potential in the apical direction. The potential changes caused by the maternal ventricular excitation are about 10 times larger than for the fetal ventricles.

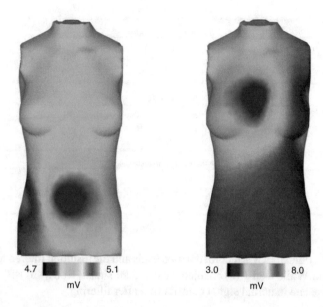

4.7 ████ 5.1   3.0 ████ 8.0
mV   mV

Fig. 2.4: Extracellular potential on the torso surface 5 ms after stimulation of the fetal (left) and maternal heart (right).

Signal processing of one clinical recording led to signals clearly representing fetal cardiac activity for some lead combinations (see Fig. 2.5). However, for other lead combinations, the maternal signal contribution is still pronounced and does not allow a clear detection of the fetal heart beats.

Fig. 2.6 illustrates one lead of the simulated ECG recording. Negative peaks, positive peaks and blunt positive hills indicate maternal depolarisation, fetal depolarisation and maternal repolarisation, respectively. The filtered signal still contains significant contributions from all of the three observed waveforms. Consequently, fetal heartbeats could not be reliably detected.

2.4 Discussion

The proposed model framework was able to produce abdominal ECGs containing maternal and fetal signal contributions. Furthermore, it was possible to simulate fetal and maternal cardiac activity in an isolated manner. This could in general be used as ground truth data for design and validation of signal processing methods. However, at the moment the simulated signals are too far from clinically observed recordings, rendering them unsuitable to tailor clinical solutions. In this study we modelled the

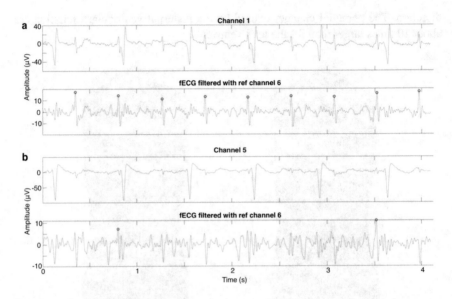

Fig. 2.5: Abdominal ECGs recorded in two leads and the resulting filtered fetal signal contribution. While lead 1 (a) enabled a clean fetal signal, lead 5 (b) still contained a lot of noise and maternal signal contribution after filtering.

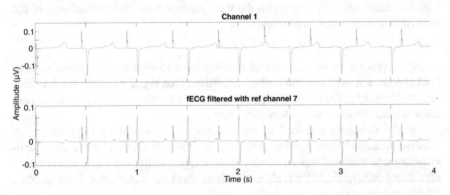

Fig. 2.6: Simulated ECG, recorded on the upper right abdomen (top) and signal extracted by the adaptive filter.

torso as a uniform medium and do not account for the varying tissue conductivities which influence the current propagation. Including additional tissue types such as bone or lungs in the model could lead to more realistic conduction and therefore more realistic fECG recordings.

Furthermore, ventricular activation was assumed to happen instantly on the whole endocardium, spreading to the epicardial surface. Several sophisticated models of ventricular activation sequences have been presented in literature [40, 41] and could be incorporated in the proposed model to enable closer resemblance of clinically observed activation patterns [42]. Future studies may also explore how time-dependent factors such as muscle contractions and respiratory movement can be incorporated.

Once a reasonable morphology of simulated signals is achieved, different artificial noise sources could be incorporated to make them more realistic and test the robustness of the proposed filtering algorithm. Besides designing signal processing algorithms, realistic simulation of fECGs could also be used to assess the influence of fetal orientation and electrode positioning. The latter could serve to establish standard electrode positions which do not yet exist.

2.5 Conclusions

We presented an electrophysiological model to simulate maternal and fetal cardiac activity during sinus rhythm. The resulting change in extracellular potential on the torso surface was used to derive the resulting voltages for several abdominal and thoracic ECG leads. Furthermore, a wavelet based adaptive filtering approach was used to extract the fetal contribution from an abdominal ECG recording for clinical as well as for simulated signals. The simulated signals contain maternal and fetal peaks but have lower complexity than clinical recordings, indicating further need to improve the proposed model.

References

1. Benoit G Bruneau. The developmental genetics of congenital heart disease. *Nature*, 451(7181):943–948, 2008.
2. Reza Sameni and Gari D Clifford. A review of fetal ECG signal processing; issues and promising directions. *The open pacing, electrophysiology & therapy journal*, 3:4, 2010.
3. Wendy N Nembhard, Elizabeth B Pathak, and Douglas D Schocken. Racial/ethnic disparities in mortality related to congenital heart defects among children and adults in the united states. *Ethn Dis*, 18(4):442–449, 2008.
4. Hilda Razzaghi, Matthew Oster, and Jennita Reefhuis. Long-term outcomes in children with congenital heart disease: National health interview survey. *The Journal of pediatrics*, 166(1):119–124, 2015.
5. AEL Van Nisselrooij, AKK Teunissen, SA Clur, L Rozendaal, E Pajkrt, IH Linskens, L Rammeloo, JMM Van Lith, NA Blom, and MC Haak. Why are congenital heart defects being missed? *Ultrasound in Obstetrics & Gynecology*, 55(6):747–757, 2020.
6. Maria Peters, John Crowe, Jean-Francois Piéri, Hendrik Quartero, Barrie Hayes-Gill, David James, Jeroen Stinstra, and Simon Shakespeare. Monitoring the fetal heart non-invasively: A review of methods. *Journal of perinatal medicine*, 29:408–16, 02 2001.
7. Paul Hamelmann, Rik Vullings, Alexander Kolen, Jan Bergmans, Judith van Laar, Piero Tortoli, and Massimo Mischi. Doppler ultrasound technology for fetal heart rate monitoring: A review. *IEEE Transactions on Ultrasonics, Ferroelectrics, and Frequency Control*, PP:1–1, 09 2019.
8. Diogo Ayres-de Campos. Electronic fetal monitoring or cardiotocography, 50 years later: what's in a name? *American Journal of Obstetrics and Gynecology*, 218:545–546, 06 2018.
9. Ella Ophir, Avshalom Strulov, Ido Solt, Rosa Michlin, Igor Buryanov, and Jacob Bornstein. Delivery mode and maternal rehospitalization. *Archives of gynecology and obstetrics*, 277:401–4, 06 2008.
10. Zarko Alfirevic, Gillian ML Gyte, Aanna Cuthbert, and Declan Devane. Continuous cardiotocography (ctg) as a form of electronic fetal monitoring (efm) for fetal assessment during labour. *Cochrane Database of Systematic Reviews*, (2), 2017.
11. L Daniel Durosier, Geoffrey Green, Izmail Batkin, Andrew J Seely, Michael G Ross, Bryan S Richardson, and Martin G Frasch. Sampling rate of heart rate variability impacts the ability to detect acidemia in ovine fetuses near-term. *Frontiers in Pediatrics*, 2:38, 2014.
12. Paul Kligfield, Leonard S Gettes, James J Bailey, Rory Childers, Barbara J Deal, E William Hancock, Gerard van Herpen, Jan A Kors, Peter Macfarlane, David M Mirvis, Olle Pahlm, Pentti Rautaharju, and Galen S Wagner. Recommendations for the standardization and interpretation of the electrocardiogram. *Journal of the American College of Cardiology*, 49(10):1109–1127, 2007.
13. Antoniya Georgieva, Patrice Abry, Václav Chudáček, Petar M Djurić, Martin G Frasch, René Kok, Christopher A Lear, Sebastiaan N Lemmens, Inês Nunes, Aris T Papageorghiou, Gerald J Quirk, Christopher WG Redman, Barry Schifrin, Jiri Spilka, Austin Ugwumadu, and Rik Vullings. Computer-based intrapartum fetal monitoring and beyond: A review of the 2nd workshop on signal processing and monitoring in labor (october 2017, oxford, uk). *Acta Obstetricia et Gynecologica Scandinavica*, 98(9):1207–1217, 2019.
14. Joachim Behar, Fernando Andreotti, Sebastian Zaunseder, Julien Oster, and Gari Clifford. A practical guide to non-invasive foetal electrocardiogram extraction and analysis. *Physiological measurement*, 37:R1–R35, 04 2016.
15. Rik Vullings and Judith OEH van Laar. Non-invasive fetal electrocardiography for intrapartum cardiotocography. *Frontiers in Pediatrics*, 8:854, 2020.
16. Abdullah Mohammed Kaleem and Rajendra D Kokate. A survey on FECG extraction using neural network and adaptive filter. *Soft Computing*, 25(6):4379–4392, 2021.
17. M Varanini, G Tartarisco, L Billeci, A Macerata, G Pioggia, and R Balocchi. An efficient unsupervised fetal QRS complex detection from abdominal maternal ECG. *Physiological Measurement*, 35(8):1607–1619, jul 2014.

18. Chen Lin, Chien-Hung Yeh, Cheng-Yen Wang, Wenbin Shi, Bess MF Serafico, Chen-Hsu Wang, Chung-Hau Juan, Hsu-Wen Vincent Young, Yenn-Jiang Lin, Hui-Ming Yeh, and Men-Tzung Lo. Robust fetal heart beat detection via r-peak intervals distribution. *IEEE Transactions on Biomedical Engineering*, 66(12):3310–3319, 2019.

19. Adam Matonia, Janusz Jezewski, Tomasz Kupka, Michał Jezewski, Krzysztof Horoba, Janusz Wrobel, Robert Czabanski, and Radana Kahankowa. Fetal electrocardiograms, direct and abdominal with reference heartbeat annotations. *Scientific Data*, 7(1):200, Jun 2020.

20. Joachim Behar, Fernando Andreotti, Sebastian Zaunseder, Qiao Li, Julien Oster, and Gari D Clifford. An ecg simulator for generating maternal-foetal activity mixtures on abdominal ecg recordings. *Physiological measurement*, 35(8):1537, 2014.

21. Hector Martinez-Navarro, Blanca Rodriguez, Alfonso Bueno-Orovio, and Ana Minchole. Repository for modelling acute myocardial ischemia: simulation scripts and torso-heart mesh. 2019.

22. Hector Martinez-Navarro, Ana Mincholé, Alfonso Bueno-Orovio, and Blanca Rodriguez. High arrhythmic risk in antero-septal acute myocardial ischemia is explained by increased transmural reentry occurrence. *Scientific reports*, 9(1):1–12, 2019.

23. Daniel D Streeter Jr, Henry M Spotnitz, Dali P Patel, John Ross Jr, and Edmund H Sonnenblick. Fiber orientation in the canine left ventricle during diastole and systole. *Circulation research*, 24(3):339–347, 1969.

24. Lazar Bibin, Jérémie Anquez, Juan Pablo de la Plata Alcalde, Tamy Boubekeur, Elsa D Angelini, and Isabelle Bloch. Whole-body pregnant woman modeling by digital geometry processing with detailed uterofetal unit based on medical images. *IEEE Transactions on Biomedical Engineering*, 57(10):2346–2358, 2010.

25. Daz 3D Studio, www.daz3d.com, accessed August 2021.

26. Sonia Dahdouh, Nadège Varsier, Antoine Serrurier, JP De la Plata, J Anquez, Elsa D Angelini, Joe Wiart, and Isabelle Bloch. A comprehensive tool for image-based generation of fetus and pregnant women mesh models for numerical dosimetry studies. *Physics in Medicine & Biology*, 59(16):4583, 2014.

27. ParaView,https://www.paraview.org, accessed August 2021.

28. Christophe Geuzaine and Jean-François Remacle. Gmsh: A 3-d finite element mesh generator with built-in pre- and post-processing facilities. *International Journal for Numerical Methods in Engineering*, 79(11):1309–1331, 2009.

29. Kirsten HWJ Ten Tusscher and Alexander V Panfilov. Alternans and spiral breakup in a human ventricular tissue model. *American Journal of Physiology-Heart and Circulatory Physiology*, 291(3):H1088–H1100, 2006.

30. Eleftheria Pervolaraki, Sam Hodgson, Arun V Holden, and Alan P Benson. Towards computational modelling of the human foetal electrocardiogram: normal sinus rhythm and congenital heart block. *Europace*, 16(5):758–765, 2014.

31. Martin J Bishop and Gernot Plank. Bidomain ecg simulations using an augmented monodomain model for the cardiac source. *IEEE transactions on biomedical engineering*, 58(8):2297–2307, 2011.

32. 2017. CARPentry, https://carpentry.medunigraz.at/, accessed August 2021.

33. Angela Agostinelli, Marla Grillo, Alessandra Biagini, Corrado Giuliani, Luca Burattini, Sandro Fioretti, Francesco Di Nardo, Stefano R Giannubilo, Andrea Ciavattini, and Laura Burattini. Noninvasive fetal electrocardiography: an overview of the signal electrophysiological meaning, recording procedures, and processing techniques. *Annals of Noninvasive Electrocardiology*, 20(4):303–313, 2015.

34. Arash Rasti-Meymandi and Aboozar Ghaffari. AECG-DecompNet: abdominal ECG signal decomposition through deep-learning model. *Physiological Measurement*, 42(4):045002, 2021.

35. Farshid Soheili Najafabadi, Edmond Zahedi, and Mohd Alauddin Mohd Ali. Fetal heart rate monitoring based on independent component analysis. *Computers in biology and Medicine*, 36(3):241–252, 2006.

36. JL Camargo-Olivares, R Martín-Clemente, S Hornillo-Mellado, MM Elena, and I Román. The maternal abdominal ECG as input to mica in the fetal ECG extraction problem. *IEEE Signal Processing Letters*, 18(3):161–164, 2011.

37. Sadasivan Puthusserypady. Extraction of fetal electrocardiogram using h(infinity) adaptive algorithms. *Medical & biological engineering & computing*, 45(10):927–937, 2007.
38. Mihaela Ungureanu, Johannes WM Bergmans, Swan Guid Oei, and Rodica Strungaru. Fetal ECG extraction during labor using an adaptive maternal beat subtraction technique. 2007.
39. Shuicai Wu, Yanni Shen, Zhuhuang Zhou, Lan Lin, Yanjun Zeng, and Xiaofeng Gao. Research of fetal ECG extraction using wavelet analysis and adaptive filtering. *Computers in Biology and Medicine*, 43(10):1622–1627, 2013.
40. Jonathan P Cranford, Thomas J O'Hara, Christopher T Villongco, Omar M Hafez, Robert C Blake, Joseph Loscalzo, Jean-Luc Fattebert, David F Richards, Xiaohua Zhang, James N Glosli, Andrew D McCulloch, David E Krummen, Felice C Lightstone, and Sergio E Wong. Efficient computational modeling of human ventricular activation and its electrocardiographic representation: A sensitivity study. *Cardiovascular Engineering and Technology*, 9(3):447–467, Sep 2018.
41. Karli Gillette, Matthias AF Gsell, Anton J Prassl, Elias Karabelas, Ursula Reiter, Gert Reiter, Thomas Grandits, Christian Payer, Darko Štern, Martin Urschler, Jason D Bayer, Christoph M Augustin, Aurel Neic, Thomas Pock, Edward J Vigmond, and Gernot Plank. A framework for the generation of digital twins of cardiac electrophysiology from clinical 12-leads ecgs. *Medical Image Analysis*, 71:102080, 2021.
42. Dirk Durrer, R Th Van Dam, GE Freud, MJ Janse, FL Meijler, and RC Arzbaecher. Total excitation of the isolated human heart. *Circulation*, 41(6):899–912, 1970.

Open Access This chapter is licensed under the terms of the Creative Commons Attribution 4.0 International License (http://creativecommons.org/licenses/by/4.0/), which permits use, sharing, adaptation, distribution and reproduction in any medium or format, as long as you give appropriate credit to the original author(s) and the source, provide a link to the Creative Commons license and indicate if changes were made.

The images or other third party material in this chapter are included in the chapter's Creative Commons license, unless indicated otherwise in a credit line to the material. If material is not included in the chapter's Creative Commons license and your intended use is not permitted by statutory regulation or exceeds the permitted use, you will need to obtain permission directly from the copyright holder.

Chapter 3
Ordinary Differential Equation-based Modeling of Cells in Human Cartilage

Kei Yamamoto[1,2], Sophie Fischer-Holzhausen[3], Maria P Fjeldstad[2], Mary M Maleckar[1]

1 – Simula Research Laboratory, Norway
2 – University of Oslo, Norway
3 – University of Bergen, Norway

Abstract Chondrocytes produce the extracellular cartilage matrix required for smooth joint mobility. As cartilage is not vascularised, and chondrocytes are not innervated by the nervous system, chondrocytes are therefore generally considered non-excitable. However, chondrocytes do express a range of ion channels, ion pumps, and receptors involved in cell homeostasis and cartilage maintenance. Dysfunction in these ion channels and pumps has been linked to degenerative disorders such as arthritis. Because the electrophysiological properties of chondrocytes are difficult to measure experimentally, mathematical modelling can instead be used to investigate the regulation of ionic currents. Such models can provide insight into the finely tuned parameters underlying fluctuations in membrane potential and cell behaviour in healthy and pathological conditions. Here, we introduce an open-source, intuitive, and extendable mathematical model of chondrocyte electrophysiology, implementing key proteins involved in regulating the membrane potential. Because of the inherent biological variability of cells and their physiological ranges of ionic concentrations, we describe a population of models that provides a robust computational representation of the biological data. This permits parameter variability in a manner mimicking biological variation, and we present a selection of parameter sets that suitably represent experimental data. Our mathematical model can be used to efficiently investigate the ionic currents underlying chondrocyte behaviour.

© The Author(s) 2022
K. J. McCabe (ed.), *Computational Physiology*, Simula SpringerBriefs on Computing 12, https://doi.org/10.1007/978-3-031-05164-7_3

3.1 Introduction

An important feature of vertebrate joints is the presence of articular cartilage, a type of connective tissue covering the end of the bones and greatly reducing the friction of bone against bone and facilitating movement. Cartilage is largely composed of proteoglycans and collagens, forming a meshwork of extracellular matrix proteins that can withstand large mechanical load while also remaining flexible compared with bone tissue [1]. As cartilage has poor regenerative properties, joint disorders like osteoarthritis are becoming an increasing health burden as the global population ages. Osteoarthritis is a progressive degradation of the cartilage in joints, and is estimated to affect 12.1% of adults over the age of 60 in the United States [2]. The progression of cartilage degradation, and what triggers it, remains poorly understood, and there are no treatments other than pain relief or joint replacement surgery [3]. Cartilage is largely made up of extracellular matrix. The cells responsible for the synthesis and turnover of this matrix are known as chondrocytes, which are embedded in small clusters in cartilaginous tissue. These cells are not innervated by the nervous system and are thus generally considered non-excitable. Additionally, as cartilage is not vascularised, they rely on diffusion in order to exchange nutrients and waste material [4]. Their extracellular milieu is therefore often slightly hypoxic, and chondrocytes have low metabolic turnover rates and poor regenerative abilities [5]. This makes them prone to degenerative disorders including osteoarthritis, in which the cartilage generally is degraded at a faster rate than it is synthesised by chondrocytes [6].

Although chondrocytes are considered non-excitable, they do express a range of ion channels and ion pumps involved in the regulation of intracellular ionic concentrations, and, in turn, the membrane potential and downstream intracellular processes. However, the exploration of electrophysiological properties of chondrocytes has been limited due to their small size (of less than $10\mu m$), representing a technical challenge for electrophysiological experiments, implying large uncertainties in measurement results [7].

A mathematical model can provide an alternative, more accessible method of studying the membrane potential and ion dynamics for chondrocytes, and the construction of the model itself can be beneficial in understanding the system's behaviour. The mathematical model we present here can be seen as a reduced and abstracted representation of what is currently known about chondrocytes. Additionally, such models can be useful to develop hypotheses, to identify mechanistic details or knowledge gaps, and to guide experiments [8]. Also, mechanistic models such as this, built on biophysical knowledge, have an increasing interest in pharmacological research as they have proven themselves useful for qualitative understanding of the underlying physiology and pathophysiology [9]. Moreover, mathematical models can find application in drug development pipelines. Strauss *et al.* [10], for instance, demonstrate in a comprehensive review the value of *in silico* models as an integrated parts of risk assessment strategies to evaluate the proarrhythmic risk of certain drugs in humans. Finally, computational models are also essential tools to study inter- and intra-variability of cell physiology. Studying the sources of variability can provide better understanding of biological processes and aid in making meaningful predictions. One intuitive approach to investigate sources of variability is to create and calibrate a population of models, which is employed in the present study [11].

Maleckar *et al.* [12] introduce a first generation mathematical model for the membrane potential of the chondrocyte. Their model is based on an intensive literature review and is additionally partly supported by experimental data. Both time-dependent and time-independent ion channel currents, as well as ion pump and ion exchanger currents, are represented in the model, with detailed description of the underlying model development and equation derivation. This first generation of the chondrocyte model mainly focused on K^+ currents across the chondrocyte cell membrane. Furthermore, Maleckar *et al.* [13] presents a model extension by including functions for the Na^+/K^+ pump, an active transport pump with an important role in setting the electrochemical gradient of Na^+ and K^+ across the cell membrane.

In this study, the model is expanded by including the ATP-sensitive K^+ (K_{ATP}) channel, a subgroup of inward-rectifying K^+ channels that have been identified in human chondrocytes [14]. This channel has been found to play an important protective role against hypoxia in the cardiovascular and nervous system [15, 16, 17], which is notable given oxygen tension in cartilage is lower than in vascularised tissue. K_{ATP} channel function is sensitive to intracellular Mg^{2+} concentrations, and channel activation is blocked by the binding of ATP [18]. Thus, when energy expenditure exceeds energy storage in the cell, intracellular ATP concentration decreases and the block of K_{ATP} channels is relieved [19], providing a direct link between cellular respiration and the membrane potential.

The previously published models were created using MATLAB [12, 13]. However, as a first-generation computational model of chondrocyte electrophysiology, we believe it beneficial to present our model on an open-source platform, so that more users can access it without having to purchase access to MATLAB. Python is an open-source programming language that has been embraced by researchers across various branches of science, owing to its powerful libraries like Numpy [20] and Scipy [21] in addition to interactive environments (IPython, Jupyter) that enable developers to provide interactive manuals of the software. In this work we present a re-implementation of the mathematical model for the chondrocyte's membrane potential fist published in 2018 by Maleckar *et al.* [12]. Our Python implementation is freely available at `https://github.com/mmaleck/chondrocyte`. By replicating figures from Maleckar *et al.* [13] we ensure a correct model transfer. The addition of K_{ATP} channel activity to the membrane dynamics of the model indicate that K_{ATP} channels may have a stabilising effect on overall K^+ currents in the presence of high Mg^{2+} concentrations.

To learn more about the model's behaviour and parameter sensitivity, a population of models was created, allowing the investigation of ranges of parameter values and their effect on the steady state of the system. This investigation identified an important role of the Na^+/K^+ pump current in the modulation of the ionic balance of the model, as well as setting the resting membrane potential. Furthermore, the model can be expanded upon by the addition of ion channel and ion pump dynamics not described here. Thus, this readily available re-implementation of a first generation chondrocyte model can provide insight into the dynamics of the chondrocyte membrane potential, and how this may be affected by variations in ion channel and ion pump function.

3.2 Methods

To investigate the chondrocyte membrane potential, we adopted a chondrocyte model and simulation protocols from previously published works [12, 13].

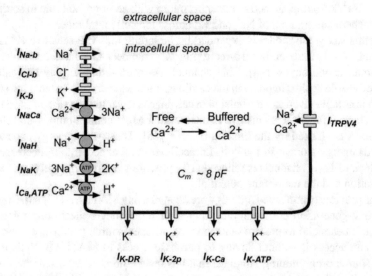

Fig. 3.1: Schematic of the model and its implemented ion channels and ion pumps. I_{Na-b}, I_{Cl-b}, and I_{K-b}: background ionic activity for Na$^+$, Cl$^-$, and K$^+$, respectively, I_{NaCa}: Na$^+$/Ca^{2+} exchanger, I_{NaH}: Na$^+$/H$^+$ exchanger, I_{NaK}: Na$^+$/K$^+$ pump, $I_{Ca,ATP}$: ATP-dependent Ca^{2+} pump, I_{K-DR}: delayed-rectifier K$^+$ channel, I_{K-2p}: two-pore K$^+$ channel, I_{K-Ca}: Ca^{2+} activated K$^+$ channel, I_{K-ATP}: ATP-sensitive K$^+$ channel, I_{TRPV4}: Transient Receptor Potential V-4 (a mechanosensitive cation channel). Adapted from [12] with permissions dictated by the Creative Commons Attribution (CC BY) license (https://creativecommons.org/licenses/by/4.0/).

The given model is defined by a set of ordinary differential equations. Those systems are defined as

$$\frac{dx}{dt} = f(x(t), u(t), k) \tag{3.1}$$

and describe the temporal evolution of states $x(t)$. In the used model, the states are the membrane potential (V_m) and the ion concentrations of Na$^+$, K$^+$, and Ca^{2+}. Furthermore, the function f depends on currents $u(f)$ for included ion pumps, channels and exchangers and model parameters k. Systems described in the form as given in Equation (3.1) need to be solved numerically. The equation which governs the chondrocyte transmembrane potential follows the Hodgkin–Huxley formalism where each cell component is represented as an electrical element, where cell membranes have a capacitance C_m and ion channels are treated as resistors. This gives a mathematical formalism for the membrane potential V_m which includes all transport processes that are electrogenic and reads as follows:

$$C_m \frac{dV_m}{dt} = - \sum_{ions} I_{ion} \qquad (3.2)$$

This work presents a model extension by including the K_{ATP} currents to the previously published chondrocyte model. Furthermore, the population of models idea is used to investigate the model behaviour under different parameter settings. The role of the maximum Na^+/K^+ pump current is studied in depth.

3.2.1 Mathematical modelling of ATP-sensitive K^+ currents

We have added an ATP-sensitive K+ channel to the model from Maleckar *et al.* [13] to investigate its role for the functionality and membrane potential of chondrocytes. The mathematical formulation of the K_{ATP} current I_{K-ATP} reads:

$$I_{K-ATP} = \sigma g_0 p_0 f_{ATP} (V_m - E_K) \qquad (3.3)$$

where σ is the channel density, g_o is the unitary channel conductance, p_0 is the maximum open channel probability, and f_{ATP} is the fraction activated channels. E_K denotes the equilibrium potential for K^+ in the given circumstances. In the previous implementation of I_{K-ATP}, a constant value was used for g_0. However, the unitary conductance can be expressed as [22]

$$g_0 = \gamma_0 f_M f_N f_T \qquad (3.4)$$

where γ_0 is the unitary conductance in the absence of intracellular Na^+ and Mg^{2+} and depends on $[K^+]_o$:

$$\gamma_0 = 35.375 \left(\frac{[K^+]_0}{5.4} \right)^{0.24} \qquad (3.5)$$

The term f_M in Equation (3.4) represents the inward rectification generated by intracellular Mg^{2+} ions and can be expressed by means of a Hill equation :

$$f_M = \frac{1}{1 + \frac{[Mg^{2+}]_i}{K_{h,Mg}}} \qquad (3.6)$$

where $K_{h,Mg}$ is the half-maximum saturation constant that depends on membrane potential and on $[K^+]_o$:

$$K_{h,Mg} = K^0_{hMg}\left([K^+]_0\right)\exp\left(-\frac{2\delta_{Mg}F}{RT}V_m\right)$$ (3.7)

$$K^0_{h,Mg}\left([K^+]_o\right) = \frac{0.65}{\sqrt{[K^+]_0+5}}$$ (3.8)

Here the value of the electrical distance δ_{Mg} was set to 0.32, F is the Faraday constant, R is the gas constant, and T is the absolute temperature. In a similar manner, the term f_N represents the inward rectification caused by intracellular Na^+ ions and is expressed as

$$f_N = \frac{1}{1+\left(\frac{[Na^+]_i}{K_{h,Na}}\right)^2}$$ (3.9)

$$K_{h,Na} = K^0_{h,Na}\exp\left(-\frac{\delta_{Na}F}{RT}V_m\right)$$ (3.10)

where $\delta_{Na} = 0.35$ and $K^0_{h,Na} = 25.9$ was used.

3.2.2 Population of Models

In the context of our work, we use the term population of models for set model simulations with randomly varied parameter sets for ionic current conductances [11]. Generally, a population of modes is useful to investigate parameter sensitivity [23, 24], sources of variability [25], as well as emergent model behavior dependency for a wide range of parameters.

In this study, we searched for parameters that had a significant effect on the simulation results and the physiological relevance of the model. Table 3.1 gives an overview of the selected set of parameters, the values in the original implementation, the parameter regime used to create a population of models, and a brief parameter description.

Parameter ranges originate from parameter sampling from log normal distribution. The parameter distribution was not fitted to experimental data and therefore underlies the assumptions that the parameters' distributions are skewed and the parameter values are positive. The distribution underlies the reasoning that numerous works have shown that a log normal distribution is often a useful assumption to describe the random variation in biological samples [26]. The population of models presented in the results is created out of 100 simulation runs. Each member of the population has its characteristic set of parameters, and all parameter set were sample from log normal distributions with the mean being the parameter value and a variance $\sigma = 0.15$.

Table 3.1: Ensemble of parameters selected for creating a population of models.

Parameter	Initial value	Selected Range	Parameter Description
Na_i	25 [mM]	17.5 - 32.5 [mM]	Na^+ initial concentration
K_i	180 [mM]	126 - 234 [mM]	K^+ initial concentration
Ca_i	10^{-5} [mM]	17.5 - 32.5 [mM]	Ca^{2+} initial concentration
$I_{NaK,scale}$	1.625	1.375 - 2.1125	scaling factor of the maximum Na^+/K^+ pump current
$g_{Na_{b,bar}}$	0.1 [pS]	0.07 - 0.13 [pS]	background Na^+ leakage conductance

3.3 Results

The following results section is divided in three parts. Firstly, we provide the validation for our new implementation that are used for tow following sections. In the second part, we present the simulations focused on the K_{ATP} currents and the effect of different Mg^{2+} concentrations. Lastly, the parameter and model behaviour is studied based on the population of models approach.

3.3.1 Validation

We performed the validation against a previous publication, Maleckar *et al.* [13]. Figure 3.2 shows the results of temperature-dependent contribution of the Na^+/K^+ pump electrogenic current to the chondrocyte resting membrane potential implemented both in MATLAB and Python as a replication of Figure 2 from Maleckar *et al.* [13]. Although our model includes other currents than shown in the figure, we conclude that our new implementation is well validated.

3.3.2 Results for the ATP-sensitive K^+ current

In order to investigate the effect of I_{K-ATP} on K^+ dynamics in the chondrocyte, numerical simulations were performed as shown in Figure 3.3. Accurate measurements of intracellular Mg^{2+} concentrations in chondrocytes are, to our knowledge, unavailable, hence we tested a range of initial Mg^{2+} values. The Mg^{2+} concentrations used for the simulation were (a) 0.1 mM (b) 1.0 mM (c) 10 mM, while three different K^+ concentrations, 5 mM, 30 mM, 70 mM, were used for each Mg^{2+} concentration as indicated in the figure. To further clarify the effect of different intracellular Mg^{2+} concentration to the overall chondrocyte matrix, we introduce Figure 3.4 where Figure 3.4 (a) shows the steady state voltage dependence of I_{K-ATP} while

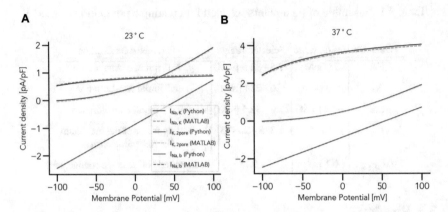

Fig. 3.2: Temperature-dependent contribution of the Na^+/K^+ pump electrogenic current to the chondrocyte resting membrane potential at **A** 23 °C and **B** 37 °C. Unbroken lines indicate our new implementation in Python, while dotted lines indicate the published MATLAB implementation.

(b) illustrates the time-dependent changes of chondrocyte membrane potential under different intracellular Mg^{2+} concentrations.

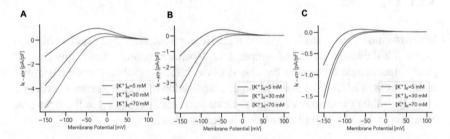

Fig. 3.3: Sensitivity of I_{K-ATP} against varying extracellular K^+ concentration based on different intracellular Mg^{2+} concentrations. Intracellular Mg^{2+} concentrations used for figures are **A** 0.1 mM **B** 1.0 mM **C** 10 mM respectively, while three concentrations of extracellular K^+ concentrations (5 mM, 30 mM, 70 mM) are used for all figures.

3.3.3 Populations of Models

A population of 100 models with randomly varied sets of parameters (Table 3.1) is illustrated in Figure 3.5. The steady state solutions for all species are reached within

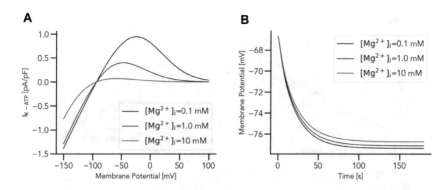

Fig. 3.4: Intracellular Mg^{2+} concentration dependent contribution of the ATP-sensitive K^+ current to the chondrocyte resting membrane potential. In **A**, we illustrate steady-state voltage dependence and the Mg^{2+} dependence of the I_{K-ATP} current, while **B** shows the hyperpolarization of chondrocyte membrane potential with different intracellular Mg^{2+} concentration. The extracellular K^+ concentration used for **A** and **B** is fixed at 7 mM.

the simulation time. Interestingly, the steady states for both Na^+ and K^+ reach similar concentration levels regardless the set of parameters, whereas Ca^{2+} concentrations at steady state solution are strongly affected by the set of parameters. It is notable that the behaviour of the membrane potential follows the shape of the Ca^{2+} time curve, and is therefore also influenced by the composition of the parameter sets.

For all 100 parameter sets, the Na^+ steady state solution is close to zero. However, the time point at which steady state is reached is parameter-dependent (Figure 3.5 **B** and **D**, Figure 3.6 **A** and **B**). To investigate the individual parameter effects, simulations with randomly drawn parameters for each individual parameter represented in Table 3.1 were performed. Larger parameter ranges were also investigated; Figure 3.5 $I_{NaK,scale}$ ranges from 1.375 to 2.1125, whereas for Figure 3.6, $I_{NaK,scale}$ values between 0.1 and 7.0 were tested. The simulation results indicate that the scaling factors of the maximum Na^+/K^+ pump current ($I_{NaK,scale}$) have a major effect on the membrane potential, the final concentrations for Na^+ and K^+, as well as the duration required for the system to reach steady-state.

A wide range of $I_{NaK,scale}$ values resulted in a Na^+ depletion, which is unlikely to occur in a physiologically viable chondrocyte preparation. A step-wise scan through $I_{NaK,scale}$ ($0.1 < I_{NaK,scale} < 7.0$) reveals three parameter regimes for the scaling of this current. For $0.1 < I_{NaK,scale} < 0.6$, here called the low parameter regime, Na^+ concentrations range from 10mM to 90mM (Figure 3.6(a)). If $0.6 < I_{NaK,scale} < 5.0$, a non-physiological state of Na^+ depletion is eventually reached, as intracellular sodium moves towards zero over time. The value of $I_{NaK,scale}$ also affects the timing of this depleted state (Figure 3.6 **B**): the smaller the scaling factor and thus the smaller the current, the later the systems reaches its steady state. Henceforth, this parameter range is referred to as the middle parameter regime. For $I_{NaK,scale} > 5.0$, steady-

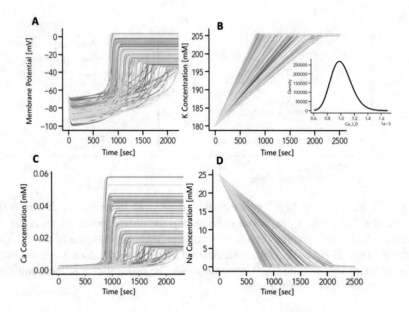

Fig. 3.5: A population of 100 models. Each simulation trajectory results from a randomly drawn parameter set. A simulation time of 2500 seconds ensures that all species reach their steady state. **A** shows the profile of membrane potential, **B** shows the profile of K$^+$ concentration, **C** shows the profile of Ca^{2+}, and **D** shows the profile of Na$^+$ concentration.

state Na$^+$ concentrations become negative, which is biologically and numerically infeasible and causes the simulation to fail.

A population of models approach was again used to further probe the model dependence on parameter variation in for low and middle parameter regimes, this time varying initial conditions (Fig. 3.7). Figure 3.7 shows strong model dependence on initial conditions for evolving variables (panels **A–D**). However, while there are a variety of stable steady states at different parameter sets, where for example, intracellular sodium is not depleted (panel **B**), all these occur for the low parameter regime of $I_{NaK,scale}$ (blue traces, Fig. 3.7, all panels). Thus, the low parameter regime of INaK conductance ($0.1 < I_{NaK,scale} < 0.6$), corresponds to, roughly, an INaK current of approximately 0.9–5.35 pA/pF and likely represents a physiologically relevant model regime incorporating a variety of stable steady states for evolving ionic concentrations and the membrane potential. To further investigate the overall dependence of the chondrocyte's resting membrane potential on K+ currents, we varied the conductance parameters for these at a variety of $I_{NaK,scale}$ values in the low parameter regime (Fig. 3.8). While there are a variety of stable steady states for the resting membrane potential of the chondrocyte in this example, even as $I_{NaK,scale}$ increases (moving right from panel **A** to **C**), all the values for each unique parameter set in the current model set the chondrocytes resting membrane potential somewhere between approximately -50 and -70 mV.

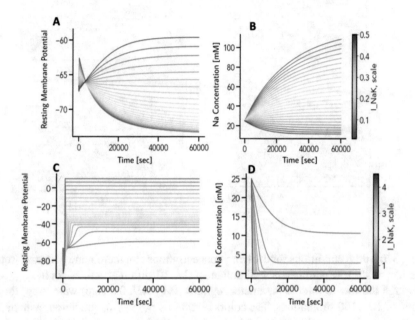

Fig. 3.6: The scaling of the maximum Na$^+$/K$^+$ pump current affects the timing and concentration of the Na$^+$ steady state as well as the membrane potential. The simulation trajectory for membrane potentials and the Na$^+$ concentrations in the low and the middle parameter regime are displayed here. Simulations for $I_{NaK,scale} > 5.0$ can not be presented because negative Na$^+$ concentrations causes the simulation to fail. **A** and **B** display simulation curves for $0.1 < I_{NaK,scale} < 0.6$. The steady state concentration of Na$^+$ ranges from 10mM up to 90mM. **C** and **D** shows simulation trajectories for $0.6 < I_{NaK,scale} < 5.0$. In all cases Na$^+$ depletion can be observed. The timing is parameter dependant.

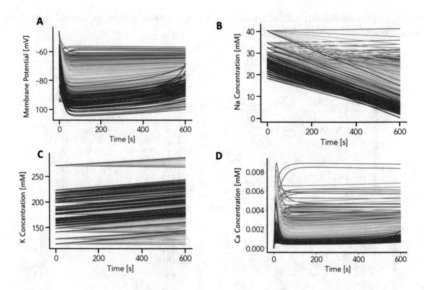

Fig. 3.7: Initial conditions for evolving concentrations and membrane potential were varied, with 100 parameters / Simultaneously, 10 different values for $I_{NaK,scale}$ through the low to middle parameter regimes (0.1 - 4.9, 0.5 step) were used, for a total of 10 x 100 simulations. The colors reveal this $I_{NaK,scale}$ gradation, with dark blue as the lowest value (0.1), and deep red as the highest value (4.9). The same colors always represent the same $I_{NaK,scale}$ value but different overall parameter sets.

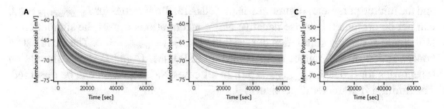

Fig. 3.8: Simulations (100 parameter sets, varying I K,b:g_K_b_bar, I K-DR: g_K_DR, I K2p: I_K_2pore_scale, I K-Ca (BK): gBK, I K, ATP: gamma_0_scale over a distribution as described in Methods) were run with differing values for $I_{NaK,scale}$ in the low parameter regime range (0.1-0.6) until intracellular sodium reached steady state. Each color in panels **A-C** above refers to a unique parameter set, and panels **A**, **B**, and **C** show the chondrocyte resting membrane potential reaching steady state over time for increasing values of $I_{NaK,scale}$ (0.1, 0.3, and 0.5, respectively).

3.4 Discussion and Conclusion

As can be seen in Figure 3.3, higher Mg^{2+} concentrations help stabilise I_{K-ATP} currents with respect to varying extracellular K^+ concentrations. Also, it can be seen from Figure 3.4 that the lower Mg^{2+} concentration contributes to the hyperpolarisation of the membrane potential. This stabilising effect is interesting, as experimental data suggest a protective role of high Mg^{2+} on joint health. Although the exact mechanisms behind this protection remain elusive, some hypotheses point to a role of Mg^{2+} in reducing chondrocyte apoptosis and facilitating chondrocyte proliferation [27, 28, 29, 30]. One caveat of this paper, however, is that physiological Mg^{2+} concentrations in chondrocytes remain unknown, and thus we have tested our model with a range of concentrations. Reliable measurements of intracellular Mg^{2+} concentrations will be necessary in order to assess the physiological effects of Mg^{2+} on K^+ currents, and the potentially protective role of Mg^{2+} and the ATP-dependent K^+ channel in cartilage maintenance.

To get a complete understanding of the three parameter regimes for the scaling factor of the maximum Na^+/K^+ pump, current further investigation is needed. It might be that the observed behaviour is a result of the chosen initial value for the Na^+ concentration. One could investigate this hypothesis by treating initial Na^+ concentrations as a parameter and scanning through a wide range of initial concentrations comparable to what has been demonstrated in Figure 3.6. But it could also be possible that the scaling parameter has this narrow parameter regime for the given model complexity and parameterisation. Additionally, it would be interesting to investigate the link between Ca^{2+} concentration and the membrane potential further.

The lack of available biological data presents a challenge for the validation of models such as the one presented here. As our study shows, measurement data are needed to determine parameter values rather than parameter ranges leading to plausible results. Due to the high parameter uncertainty of the model, care must be taken in interpreting results in order to make useful predictions and to suggest hypotheses for further testing. A more thorough investigation into the range of appropriate ionic concentrations remains to be performed.

Here we present a model of chondrocyte electrophysiology, with particular attention to the role of K^+ currents in setting the steady-state membrane potential. Our preliminary results support a protective role of Mg^{2+} in cartilage maintenance as recorded in clinical studies by potentially stabilising K^+ currents through the ATP-dependent K^+ channel, although more research is required. Furthermore, we show how a population of models can be used to examine fluctuations in a range of parameters, and their interactions as the model reaches steady-state membrane potential. Such random variations in parameter values will also act to make the model more realistic and more robust to naturally occurring fluctuations as seen in biological data. Finally, the chondrocyte model has been re-implemented from MATLAB into Python to increase its accessibility, and is available as an open-source repository on Github, with demo scripts to aid interested parties in getting started.

References

1. Savio Lau-Yuen Woo and Joseph Addision Buckwalter. Injury and repair of the musculoskeletal soft tissues. *Journal of Orthopaedic Research*, 6(6):907–931, 1988.
2. Charles F Dillon, Elizabeth K Rasch, Qiuping Gu, and Rosemarie Hirsch. Prevalence of knee osteoarthritis in the united states: Arthritis data from the third national health and nutrition examination survey 1991-94. *The Journal of Rheumatology*, 11(33):2271–2279, 2006.
3. Susanne Grässel and Dominique Muschter. Recent advances in the treatment of osteoarthritis [version 1; peer review: 3 approved]. *F1000Research*, 325(9(F1000 Faculty Rev)):1–17, 2020.
4. Charles W Archer and Philippa Francis-West. The chondrocyte. *The International Journal of Biochemistry Cell Biology*, 4(35):401–404, 2003.
5. Ramesh Rajpurohit, Cameron J Koch, Zhuliang Tao, Cristina Maria Teixeira, and Irving M Shapiro. Adaptation of chondrocytes to low oxygen tension: Relationship between hypoxia and cellular metabolism. *Journal of Cellular Physiology*, 2(168):424–432, 1996.
6. Hemanth Akkiraju and Anja Nohe. Role of chondrocytes in cartilage formation, progression of osteoarthritis and cartilage regeneration. *Journal of Developmental Biology*, 4(3):177–192, 2015.
7. James R Wilson, Robert B Clark, Umberto Banderali, and Wayne R Giles. Measurement of the membrane potential in small cells using patch clamp methods. *Channels (Austin)*, 5(6):530–537, 2011.
8. Olaf Wolkenhauer. Why model? *Frontiers in physiology*, 5:21, 2014.
9. Piet H van der Graaf, Neil Benson, and Lambertus A Peletier. Topics in mathematical pharmacology. *Journal of Dynamics and Differential Equations*, 28(3):1337–1356, 2016.
10. David G Strauss, Wendy W Wu, Zhihua Li, John Koerner, and Christine Garnett. Translational models and tools to reduce clinical trials and improve regulatory decision making for qtc and proarrhythmia risk (ich e14/s7b updates). *Clinical Pharmacology & Therapeutics*, 109(2):319–333, 2021.
11. Oliver J Britton, Alfonso Bueno-Orovio, Karel Van Ammel, Hua Rong Lu, Rob Towart, David J Gallacher, and Blanca Rodriguez. Experimentally calibrated population of models predicts and explains intersubject variability in cardiac cellular electrophysiology. *Proceedings of the National Academy of Sciences*, 110(23):E2098–E2105, 2013.
12. Mary M Maleckar, Robert B Clark, Bartholomew Votta, and Wayne R Giles. The resting potential and K+ currents in Primary Human articular chondrocytes. *Frontiers in Physiology*, 9(SEP):1–21, 2018.
13. Mary M Maleckar, Pablo Martín-Vasallo, Wayne R Giles, and Ali Mobasheri. Physiological Effects of the Electrogenic Current Generated by the Na + /K + Pump in Mammalian Articular Chondrocytes . *Bioelectricity*, 2(3):258–268, 2020.
14. A Mobasheri, TC Gent, AI Nash, MD Womack, CA Moskaluk, and R Barrett-Jolley. Evidence for functional atp-sensitive (k(atp)) potassium channels in human and equine articular chondrocytes. *Osteoarthritis and Cartilage*, 15(1):1–8, 2007.
15. Katsuya Yamada, Juan Juan Ji, Hongjie Yuan, Takashi Miki, Shininchi Sato, Naoki Horimoto, Tetsuo Shimizu, Susumu Seino, and Nobuya Inagaki. Protective role of atp-sensitive potassium channels in hypoxia-induced generalized seizure. *Science*, 29(5521):1543–156, 2001.
16. Russel M Crawford *et al.* Chronic mild hypoxia protects heart-derived h9c2 cells against acute hypoxia/reoxygenation by regulating expression of the sur2a subunit of the atp-sensitive k+ channel. *Membrane Transport, Structure, Function, and Biogenesis*, 278(33):31444–31455, 2003.
17. Maurizio Turzo, Julian Vaith, Felix Lasitschka, Markus A Weigand, and Cornelius J Busch. Role of atp-sensitive potassium channels on hypoxic pulmonary vasoconstriction in endotoxemia. *Respiratory Research*, 19(29), 2018.
18. S Trapp, SJ Tucker, and FM Ashcroft. Activation and inhibition of k-atp currents by guanine nucleotides is mediated by different channel subunits. *Proc Natl Acad Sci USA*, 16(94):8872–8877, 1997.
19. A Noma. Atp-regulated k+ channels in cardiac muscle. *Nature*, 305:147–148, 1983.

20. Charles R Harris *et al.* Array programming with NumPy. *Nature*, 585(7825):357–362, 2020.
21. Pauli Virtanen *et al.* SciPy 1.0: fundamental algorithms for scientific computing in Python. *Nature Methods*, 17(3):261–272, 2020.
22. José M Ferrero, Javier Sáiz, José M Ferrero, and Nitish V Thakor. Simulation of action potentials from metabolically impaired cardiac myocytes: Role of ATP-sensitive K+ current. *Circulation Research*, 79(2):208–221, 1996.
23. Eve Marder and Adam L Taylor. Multiple models to capture the variability in biological neurons and networks. *Nature neuroscience*, 14(2):133–138, 2011.
24. Márcia R Vagos, Hermenegild Arevalo, Bernardo Lino de Oliveira, Joakim Sundnes, and Mary M Maleckar. A computational framework for testing arrhythmia marker sensitivities to model parameters in functionally calibrated populations of atrial cells. *Chaos: An Interdisciplinary Journal of Nonlinear Science*, 27(9):093941, 2017.
25. Diane R Mould and Richard Neil Upton. Basic concepts in population modeling, simulation, and model-based drug development—part 2: introduction to pharmacokinetic modeling methods. *CPT: pharmacometrics & systems pharmacology*, 2(4):1–14, 2013.
26. Eckhard Limpert, Werner A Stahel, and Markus Abbt. Log-normal distributions across the sciences: keys and clues: on the charms of statistics, and how mechanical models resembling gambling machines offer a link to a handy way to characterize log-normal distributions, which can provide deeper insight into variability and probability—normal or log-normal: that is the question. *BioScience*, 51(5):341–352, 2001.
27. Frank Feyerabend, Frank Witte, Michael Kammal, and Regine Willumeit. Unphysiologically high magnesium concentrations support chondrocyte proliferation and redifferentiation. *Tissue Engineering*, 12(12):3545–3556, 2006.
28. Sijing Li, Fenbo Ma, Xiangchao Pang, Bin Tang, and Lijun Lin. Synthesis of chondroitin sulfate magnesium for osteoarthritis treatment. *Carbohydrate Polymers*, 212:387–394, 2019.
29. H Yao, JK Xu, NY Zheng, JL Wang, SW Mok, YW Lee, L Shi, JY Wang, J Yue, SH Yung, PJ Hu, YC Ruan, YF Zhang, KW Ho, and L Qin. Intra-articular injection of magnesium chloride attenuates osteoarthritis progression in rats. *Osteoarthritis and Cartilage*, 27(12):1811–1821, 2019.
30. Xiaoqing Kuang, Jiachi Chiou, Kenneth Lo, and Chunyi WEN. Magnesium in joint health and osteoarthritis. *Nutrition Research*, 90:24–35, 2021.

Open Access This chapter is licensed under the terms of the Creative Commons Attribution 4.0 International License (http://creativecommons.org/licenses/by/4.0/), which permits use, sharing, adaptation, distribution and reproduction in any medium or format, as long as you give appropriate credit to the original author(s) and the source, provide a link to the Creative Commons license and indicate if changes were made.

The images or other third party material in this chapter are included in the chapter's Creative Commons license, unless indicated otherwise in a credit line to the material. If material is not included in the chapter's Creative Commons license and your intended use is not permitted by statutory regulation or exceeds the permitted use, you will need to obtain permission directly from the copyright holder.

Chapter 4
Conduction Velocity in Cardiac Tissue as Function of Ion Channel Conductance and Distribution

Kristian Gregorius Hustad[1], Ena Ivanovic[2], Adrian Llop Recha[3], Abinaya Abbi Sakthivel[3]

1 – Simula Research Laboratory, Norway
2 – Dept. of Physiology, University of Bern, Switzerland
3 – Dept. of Informatics, University of Oslo, Norway

Abstract Ion channels on the membrane of cardiomyocytes regulate the propagation of action potentials from cell to cell and hence are essential for the proper function of the heart. Through computer simulations with the classical monodomain model for cardiac tissue and the more recent extracellular-membrane-intracellular (EMI) model where individual cells are explicitly represented, we investigated how conduction velocity (CV) in cardiac tissue depends on the strength of various ion currents as well as on the spatial distribution of the ion channels. Our simulations show a sharp decrease in CV when reducing the strength of the sodium (Na^+) currents, whereas independent reductions in the potassium (K1 and Kr) and L-type calcium currents have negligible effect on the CV. Furthermore, we find that an increase in number density of Na^+ channels towards the cell ends increases the CV, whereas a higher number density of K1 channels slightly reduces the CV. These findings contribute to the understanding of ion channels (e.g. Na^+ and K^+ channels) in the propagation velocity of action potentials in the heart.

4.1 Introduction

A healthy heart rhythm is essential for the proper functioning of the cardiac pump, and requires the coordinated propagation of electrical impulses through the myocardium. The cardiac action potential is a change in the membrane potential governed by the ionic current flowing through ion channels, which are distributed along the cell membrane. Current flowing into the cell through activated sodium (Na^+) channels is responsible for the rapid upstroke of the action potential [1]. This is followed by

© The Author(s) 2022
K. J. McCabe (ed.), *Computational Physiology*, Simula SpringerBriefs
on Computing 12, https://doi.org/10.1007/978-3-031-05164-7_4

a current inflow through L-type calcium (CaL) channels and an outflow through different types of potassium channels (e.g. Kr and K1) leading to repolarisation, thus bringing the cell membrane to resting membrane potential [1]. More precisely, I_{Kr} is the rapid component of the delayed rectifier, and I_{K1}, the inward rectifying potassium current, which stabilises the resting membrane potential [1]. L-type calcium channels, on the other hand, are also responsible for the excitation-contraction coupling of the cardiac muscle [1].

Disorders of electrical conduction, such as slow conduction and conduction block, can lead to life-threatening arrhythmias, which occur frequently in the diseased heart. In this study, we investigated how both the strength of several transmembrane ionic currents and the spatial distribution of ion channels along the cell membrane influence conduction velocity (CV) in cardiac tissue. Our aim is to compare the effect on CV with the use of two computational models: the monodomain model and the extracellular-membrane-intracellular (EMI) model.

The monodomain model is a classical approximation of the electrical propagation in myocardial tissue based on a homogenised mathematical model of the cell. The intra- and extracellular domains overlap and are considered continuous. As a consequence, the monodomain model offers a good insight of large scale effects, but it is limited when sizes are reduced to a single cell. Alternatively, the EMI model represents the extracellular, the membrane, and the intracellular spaces at the expense of computational power. Therefore, one of the main advantages is the possibility to introduce changes in local cell properties (e.g. changes in ion channel density distribution on the cell membrane) that might contribute significantly to action potential propagation [2], [3].

4.2 Models and methods

4.2.1 The monodomain model

The monodomain model is a simplification of the bidomain model [4]. In the bidomain model, heart tissue is classified into two groups or domains: extracellular and intracellular, defined by their respective electric potentials, u_e and u_i, and conductivities G_e and G_i.

Each point in the heart is considered to be in both domains. Therefore, both spaces overlap.

The physical description can be addressed using a generalisation of Ohm's Law. The current densities at each domain will be: $J_i = -G_i \nabla u_i$, and $J_e = -G_e \nabla u_e$. Assuming that there are no other sources than the membrane, the conservation of charge applies, and thus: $\nabla(J_i + J_e) = 0$.

The current flowing from one domain to the other through the cell membrane is called transmembrane current, I_m. Because the charge is conserved, then $\nabla J_e = -\nabla J_i = I_m$. Transmembrane current (Eq. 4.1) depends on the voltage drop between both domains, $v = u_i - u_e$, the membrane capacitance C_m, the ionic current, I_{ion}, and surface area-to-volume ratio of cardiac cell, β_m.

$$I_m = \beta_m \left(C_m \frac{\partial v}{\partial t} + I_{\text{ion}}(v) \right) \tag{4.1}$$

It can be shown that the set of equations governing the bidomain model are the ones expressed in Eq.4.2 and Eq.4.3.

$$\nabla G_i (\nabla v + \nabla u_e) = \beta_m \left(C_m \frac{\partial v}{\partial t} + I_{\text{ion}}(v) \right) \tag{4.2}$$

$$\nabla G_i \nabla v + \nabla (G_i + G_e) \nabla u_e = 0 \tag{4.3}$$

Solving the bidomain equation is a computationally heavy process. Therefore, the monodomain model is often used instead. In the monodomain model, the anisotropy between extra- and intracellular spaces is assumed to be the same, i.e., their respective conductances are proportional $G_i = \lambda G_e$.

If we define an effective conductivity, $G_{\text{eff}} = \frac{\lambda}{1+\lambda} G_i$, then the bidomain equation can be simplified and rearranged as shown in Eq.4.4.

$$\frac{\partial v}{\partial t} = \frac{1}{C_m \beta_m} \nabla G_{\text{eff}} \nabla v - \frac{1}{C_m} I_{\text{ion}} \tag{4.4}$$

4.2.2 The EMI model

In the EMI model [5], the extracellular (E), cell membrane (M) and intracellular (I) domains, are represented explicitly as depicted in Figure 4.1. The intracellular spaces of both cells, Ω_i^1 and Ω_i^2, are separated from the extracellular domain, Ω_e, by the cell membrane boundaries, Γ_1 and Γ_2. Additionally, $\Gamma_{1,2}$ is the boundary separating the intracellular domains of two connected cells. Note that the EMI model is always solved in a three-dimensional space, as the extracellular space should be a single, connected domain.

The equation system describing the potentials in the EMI model is summarised in the following set of equations:

$$\nabla \cdot \sigma_e \nabla u_e = 0 \quad \text{in } \Omega_e, \qquad n_e \cdot \sigma_e \nabla u_e = -n_i^k \cdot \sigma_i \nabla u_i^k \equiv I_m^k \quad \text{at } \Gamma_k,$$

$$\nabla \cdot \sigma_i \nabla u_i^k = 0 \quad \text{in } \Omega_i^k, \qquad v^k = \frac{1}{C_m} (I_m^k - I_{\text{ion}}^k) \quad \text{at } \Gamma_k,$$

$$u_e = 0 \quad \text{at } \partial\Omega_e^D, \qquad n_i^2 \cdot \sigma_i \nabla u_i^2 = -n_i^1 \cdot \sigma_i \nabla u_i^1 \equiv I_{1,2} \quad \text{at } \Gamma_{1,2},$$

$$n_e \cdot \sigma_e \nabla u_e = 0 \quad \text{at } \partial\Omega_e^N, \qquad u_i^1 - u_i^2 = w \quad \text{at } \Gamma_{1,2},$$

$$u_i^k - u_e = v^k \quad \text{at } \Gamma_k,$$

$$s_t^k = F^k \quad \text{at } \Gamma_k, \qquad w = \frac{1}{C_{1,2}} (I_{1,2} - I_g^k) \quad \text{at } \Gamma_{1,2},$$

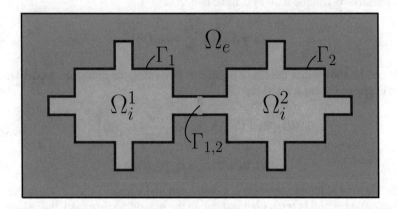

Fig. 4.1: Two-dimensional schematic of the different domains for two connected cells in the EMI model. Adapted from [5] with permissions dictated by the Creative Commons CC BY license (https://creativecommons.org/licenses/by/4.0/).

As in the case of the monodomain model, potentials of intracellular, extracellular and transmembrane domains are denoted by u_i, u_e and v, respectively. In addition to the ion current, I_{ion}, and the transmembrane current, I_m, defined at the cell membrane, Γ_k, a gap junction current, I_g, and a transmembrane current, $I_{1,2}$, are defined at the interface between two cells, $\Gamma_{1,2}$, in order to model the gap junction dynamics.

The gap junction between neighbouring cells, $\Gamma_{1,2}$, is modeled as a passive membrane with a constant resistance, R_g. Its electric dynamics are described using the currents I_g and $I_{1,2}$, and the potential drop at the junction, w.

Furthermore, C_m and $C_{1,2}$ are the transmembrane and the gap junction capacitances. σ_i and σ_e are the conductivities of both intracellular and extracellular domain, whereas n_i and n_e are the outward pointing normal vectors of the inner and outer cell domains, respectively.

The boundary of the extracellular domain, $\partial\Omega_e$, is further divided into two parts: one corresponding to the Dirichlet boundary conditions, $\partial\Omega_e^D$, and other one corresponding to the Neumann boundary conditions, $\partial\Omega_e^N$. Additionally, the index k can take the values 1 or 2 depending on which cell is described. Two cells were used to introduce the EMI model, however the model can easily be scaled up to consider more cells in the system.

Finally, s represents a collection of additional state variables introduced in the membrane model, whereas $F(v,s)$ represents the ordinary differential equations describing the dynamics of the additional state variables.

Since there is no analytical solution to the EMI model, a numerical solution is required (see [5, 3] for a discussion of numerical methods for the EMI model).

4.3 Results

We use code based on [6] to solve the monodomain and EMI models using finite difference method discretisation, and study the influence on CV. The simulations are run over a domain size of $2000\,\mu m \times 40\,\mu m$ for the monodomain model and $1956\,\mu m \times 40\,\mu m \times 30\,\mu m$ for the EMI model, both with a spatial resolution of $\Delta x = \Delta y = \Delta z = 2\,\mu m$. For the EMI model the cells are arranged in a line such that all cells are connected in the x direction, similar to Figure 4.1. Each cell is comprised of five disjoint subdomains, as depicted in Figure 4.3, and their extents are listed in Table 4.2. Regarding the time domain, the system evolves during 5 ms in steps of 0.01 ms.

The base model representing the cell membrane is described in [7]. Furthermore, the values of the most relevant parameters used in both monodomain and EMI model simulations are compiled in Table 4.1 and Table 4.2, respectively.

Parameter	Value
G_i (x direction)	$2.9\,\mathrm{mS\,cm^{-1}}$
G_i (y direction)	$1.0\,\mathrm{mS\,cm^{-1}}$
λ	$2/3$
C_m	$1.0\,\mathrm{\mu F\,cm^{-2}}$
β_m	$2000\,\mathrm{cm^{-1}}$

Table 4.1: Relevant parameter values used in the monodomain simulations.

Parameter	Value	Domain	Extent
C_m	$1.0\,\mathrm{\mu F\,cm^{-2}}$	Ω_O	$100\,\mu m \times 20\,\mu m \times 20\,\mu m$
$C_{1,2}$	$1.0\,\mathrm{\mu F\,cm^{-2}}$	Ω_N	$16\,\mu m \times 4\,\mu m \times 16\,\mu m$
σ_e	$20.0\,\mathrm{mS\,cm^{-1}}$	Ω_S	$16\,\mu m \times 4\,\mu m \times 16\,\mu m$
σ_i	$4.0\,\mathrm{mS\,cm^{-1}}$	Ω_W	$4\,\mu m \times 16\,\mu m \times 16\,\mu m$
R_g	$0.0045\,\mathrm{k\Omega\,cm^2}$	Ω_E	$4\,\mu m \times 16\,\mu m \times 16\,\mu m$

Table 4.2: Relevant parameter values used in the EMI simulations.

Our study aims to investigate the CV dependence with ion channel properties from two different perspectives: when ion channel conductances change, and when ion channel distributions along the cell membrane is modified.

First, for exploring the relation between CV and ion channel conductance, we focused particularly on the Na$^+$, K1, Kr and CaL channels. The nominal values for the Na, K1 and Kr channel conductances are 12.6, 0.37 and $0.025\,\mathrm{mS\,\mu F^{-1}}$, respectively, while the nominal value for CaL channel conductance is $0.12\,\mathrm{nL\,\mu F^{-1}\,ms^{-1}}$ as specified in [7]. Every conductance was varied from 20% to 150% of their respective nominal value by sweeping an adjustment factor from 0.2 to 1.5 in steps of 0.1.

Figure 4.2 shows the resultant CV dependence with each channel conductance in both models, the monodomain model and the EMI model.

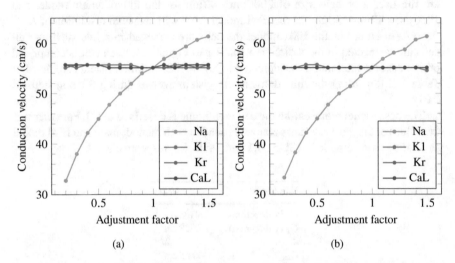

(a)   (b)

Fig. 4.2: CV dependence with channel conductance of Na, CaL, K1 and Kr channels simulated using (a) the monodomain and (b) EMI models.

In order to address the second part of our study, we change the uniform distribution of ion channels into a non-uniform distribution along the cell membranes. Changes in local properties of cells can only be implemented using the EMI model by allowing movement of ion channels towards the cell ends, i.e., the region that is closest to the $\Gamma_{1,2}$ domain (see Figure 4.1). To run the simulations, we considered two types of channels, Na^+ and K1, and we explored their four corner distribution states, i.e, when both Na^+ and K1 are uniformly distributed, when both type of channels are completely shifted towards the cell end (see Figure 4.3), and a combination of these two. The resulting CV for each case is compiled in Table 4.3.

	Uniform K1	Non-uniform K1
Uniform Na	55.1 cm/s	54.0 cm/s
Non-uniform Na	60.0 cm/s	58.7 cm/s

Table 4.3: CV (cm/s) with uniform and non-uniform channel distribution.

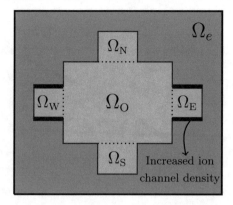

Fig. 4.3: In our simulations of a non-uniform distribution of ion channels, the ion channel density was increased in the brown areas near the cell ends (Ω_W and Ω_E) and decreased elsewhere (Ω_N, Ω_S and Ω_O).

4.4 Discussion

4.4.1 Influence of ion channel conductance on CV

Figure 4.2 shows that the CV increases monotonously as a function of the sodium channel conductance, g_{Na}, and this dependency is observed for both models.

To assess the compatibility between the monodomain and EMI model, the calculated CV points for Na^+ in Figure 4.2 are fitted into a function of the form $CV(g_{Na}) = a \cdot g_{Na}^p$, where a and p are constants, using a non-linear least-squares method. The curve adjustment is shown in Figure 4.4. For the constant p, we obtain $p = 0.294$ and $p = 0.3$ for the EMI model and the monodomain model, respectively. Thus, consistent results were obtained with both models.

Furthermore, Figure 4.2 shows that CV remains almost constant when sweeping K^+ and CaL channels conductances. Therefore, varying the strength of these ion channels did not lead to significant changes in CV.

4.4.2 Influence of ion channel distribution

From Table 4.3, when both K1 and Na^+ channels are uniformly distributed, the CV reaches 55.1 cm/s. However, when all Na^+ channels are placed at the cell ends (the coupling junction area between neighbouring cells) while keeping K1 channels uniformly distributed, then the CV experienced an increment of around

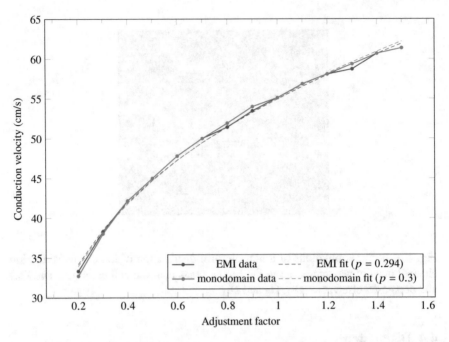

Fig. 4.4: Fitting the CV to the function $CV(g_{\mathrm{Na}}) = a \cdot g_{\mathrm{Na}}^{p}$.

9% with respect to the previous situation. Conversely, when K1 channels are moved towards the cell ends and $\mathrm{Na^+}$ channels are kept uniformly distributed along the cell membrane, CV decreased around 2%.

These results suggest that $\mathrm{Na^+}$ channel distribution contributes significantly to the overall CV, whereas the effect of K1 distribution is relatively small. While $\mathrm{Na^+}$ channels increased the CV as their accumulation to the cell end increased, the opposite effect was observed with a non-uniform distribution of K1 channels. Additionally, we deduce that both contributions are asymmetrical. The displacement of $\mathrm{Na^+}$ channels causes a major impact on CV compared to the displacement of K1 channels. When both channels are non-uniformly distributed, the CV increases about 6.5% (see Table 4.3). This CV is the result of the displacement of $\mathrm{Na^+}$ channels, which largely increases CV, mitigated by the displacement of K1 channels, which slightly decreases CV.

Immunohistochemical studies revealed that about 50% of the $\mathrm{Na^+}$ channels are located in the membranes of the intercalated discs [8]. In the diseased heart with reduced gap junctional coupling, action potential propagation can be maintained through a mechanism known as ephaptic coupling. A prerequisite for ephaptic effects to occur is a high density of $\mathrm{Na^+}$ channels at the intercalated disc, where the intermembrane distance between two adjacent cells is small (< 30 nm, [9]).

4.5 Conclusions

In this study, we investigated the influence of ion channels, their conductance and their physical distribution along the cell membrane, on CV. To that end, two different models were used: the monodomain model and the EMI model. While the former offers a good insight of large scale effects by reducing the cell model complexity, the latter allows the implementation of changes in local cell properties, at the expense of increased computational effort.

Regarding the CV dependence on ion channel conductance, the study focused particularly on Na^+, K^+ (K1 and Kr) and CaL channels. Both models showed that Na^+ channel conductance strongly influences CV, whereas the effect of the other ion channel conductances on CV is negligible. Moreover, the changes in CV as a result of modifying Na^+ and K1 channels distribution on the cell membranes, were explored with the use of the EMI model. The simulation results suggest that the influence of these channels to the CV is opposed and asymmetrical. The influence is considered opposed because the CV increases, when Na^+ channels are moved towards the cell ends, but decreases in the case of K1 channels being located at the cell ends. Furthermore, the effect on CV is asymmetrical because the movement of Na^+ channels along the cell membrane causes a substantial modification of the CV, of around 9%, compared to the transfer of K1 channels, which accounts for a variation of 2%.

Acknowledgements The illustrations in Figure 4.1 and Figure 4.3 were created by Karoline Horgmo Jæger and reused with permission. The authors are grateful to Jæger for proofreading the manuscript and for providing code solving the monodomain and EMI models, and to Aslak Tveito for discussions about our findings in this study.

References

1. Bertil Hille. *Ionic channels of excitable membranes*. Sinauer Associates, Sunderland, MA, U.S.A., 1992.
2. RH Clayton, O Bernus, EM Cherry, H Dierckx, FH Fenton, L Mirabella, AV Panfilov, FB Sachse, G Seemann, and H Zhang. Models of cardiac tissue electrophysiology: Progress, challenges and open questions. *Progress in Biophysics and Molecular Biology*, 104(1):22–48, 2011. Cardiac Physiome project: Mathematical and Modelling Foundations.
3. Karoline Horgmo Jæger, Kristian Gregorius Hustad, Xing Cai, and Aslak Tveito. Efficient numerical solution of the emi model representing the extracellular space (e), cell membrane (m) and intracellular space (i) of a collection of cardiac cells. *Frontiers in Physics*, 8:539, 2021.
4. L Tung. *A bidomain model for describing ischemic myocardial D-C potentials*. PhD thesis, M.I.T. Cambridge, Mass., 1978.
5. Aslak Tveito, Karoline Horgmo Jæger, Miroslav Kuchta, Kent-Andre Mardal, and Marie E Rognes. A cell-based framework for numerical modeling of electrical conduction in cardiac tissue. *Frontiers in Physics*, 5:48, 2017.
6. Karoline Horgmo Jæger, Andrew G Edwards, Andrew McCulloch, and Aslak Tveito. Properties of cardiac conduction in a cell-based computational model. *PLOS Computational Biology*, 15(5):1–35, 05 2019.
7. Karoline Horgmo Jæger, Verena Charwat, Bérénice Charrez, Henrik Finsberg, Mary M Maleckar, Samuel Wall, Kevin E Healy, and Aslak Tveito. Improved computational identification of drug response using optical measurements of human stem cell derived cardiomyocytes in microphysiological systems. *Frontiers in Pharmacology*, 10:1648, 2020.
8. SA Cohen. Immunocytochemical localization of rh1 sodium channel in adult rat heart atria and ventricle. presence in terminal intercalated disks. *Circulation*, 94:3083–3086, 1996.
9. Y Mori, GI Fishman, and CS Peskin. Ephaptic conduction in a cardiac strand model with 3d electrodiffusion. *Proc Natl Acad Sci U S A*, 105:6463–6468, 2008.

Open Access This chapter is licensed under the terms of the Creative Commons Attribution 4.0 International License (http://creativecommons.org/licenses/by/4.0/), which permits use, sharing, adaptation, distribution and reproduction in any medium or format, as long as you give appropriate credit to the original author(s) and the source, provide a link to the Creative Commons license and indicate if changes were made.

The images or other third party material in this chapter are included in the chapter's Creative Commons license, unless indicated otherwise in a credit line to the material. If material is not included in the chapter's Creative Commons license and your intended use is not permitted by statutory regulation or exceeds the permitted use, you will need to obtain permission directly from the copyright holder.

Chapter 5
Computational Prediction of Cardiac Electropharmacology - How Much Does the Model Matter?

John Dawson[1], Anna Gams[2], Ivan Rajen[3], Andrew M Soltisz[4], Andrew G Edwards[5]

1 – Dept. of Physiology and Membrane Biology, University of California Davis, USA
2 – Dept. of Biomedical Engineering, The George Washington University, USA
3 – Dept. of Bioengineering, University of California San Diego, USA
4 – Dept. of Biomedical Engineering, The Ohio State University, USA
5 – Simula Research Laboratory, Norway

Abstract Animal data describing drug interactions in cardiac tissue are abundant, however, nuanced inter-species differences hamper the use of these data to predict drug responses in humans. There are many computational models of cardiomyocyte electrophysiology that facilitate this translation, yet it is unclear whether fundamental differences in their mathematical formalisms significantly impact their predictive power. A common solution to this problem is to perform inter-species translations within a collection of models with internally consistent formalisms, termed a "lineage", but there has been little effort to translate outputs across lineages. Here, we translate model outputs between lineages from Simula and Washington University for models of ventricular cardiomyocyte electrophysiology of humans, canines, and guinea pigs. For each lineage-species combination, we generated a population of 1000 models by varying common parameters, namely ion conductances, according to a Guassian log-normal distribution with a mean at the parameter's species-specific default value and standard deviation of 30%. We used partial least squares regression to translate the influences of one model to another using perturbations to calculated descriptors of resulting electrophysiological behavior derived from these parameter variations. Finally, we evaluated translation fidelity by performing a sensitivity analysis between input parameters and output descriptors, as similar sensitivities between models of a common species indicates similar biological mechanisms underlying model behavior. Successful translation between models, especially those from different lineages, will increase confidence in their predictive power.

5.1 Introduction

Preclinical drug development relies on animal-derived data for determining drug efficacy and toxicity. This is especially the case for interactions within cardiac physiology, yet there is a crisis in translating these data to human outcomes and clinical utility or liability [1]. This crisis is consequential: one third of all discontinued drugs are withdrawn from the market on account of poor safety, with cardiovascular toxicity being the leading cause of both post-approval and preclinical withdrawal [2]. In other words, cardiotoxic reactions as predicted in animal models are preventing a significant portion of drugs from advancing to human trials, and drugs that manage to pass approval are having unanticipated cardiotoxicity in humans. To both understand the differences between non-human preclinical species as experimental models, and to better translate those animal-based screening results to human outcomes, many groups have developed computational models that recapitulate the cardiomyocyte electrophysiology of each species [3, 4, 5, 6]. These models are an effective and inexpensive way to predict mechanisms of action for drug activity and toxicities and have significant utility for human medicine.

However, there are sources of variability that hamper translating discoveries made in one computational model to another. In the case of inter-species translation, the most obvious are the significant and characteristic phenotypic differences between species that reflect the meaningful and real differences in their cardiac electrophysiology. Additionally, the mathematical formulations used to describe ion channel open probabilities, gating mechanisms, or other model components chosen by the model's developers, may differ between models of the same species. Laboratories often develop a computational model for one species that uses their previous formulations to parameterize models for another species. This has created collections of models sharing common mathematical formulations, but for which parameters are varied in order to capture key physiologic differences among different species. These distinct collections are often termed "lineages", and it is unclear whether their differing formulations impact cross-lineage predictions. Finally, all existing models can only represent an idealized phenotype of their respective species, further complicating accurate model-based species-translations. That is, they capture the functional properties of the average cardiomyocyte of that species. Inherent biological heterogeneity and individual variability reflected in experimental measurements are not represented within the single set of parameter values used in a computational model. This valuable biological information is unused and needs to be accounted for when developing and effectively translating results between these models. This is particularly true for cardiac safety screening, where rare but lethal events are a critical outcome. Generating populations of models whose parameters vary in a way that represents both intra- and inter-individual variability is one potential method for overcoming this issue.

Prior work has attempted to translate simulated drug effects between computational models of cardiac electrophysiology, with varied success, but such work has typically only been conducted within a single model lineage or using an idealized model representing average behavior. For example, Tveito *et al.* demonstrated methods for translating drug effects from animal experiments to computational models of ventricular myocytes for a panel of pro-arrhythmic drugs [7]. Population-based modeling has recently been developed to overcome this drawback by generating sets of models from which statistical analyses can be harnessed. For example, by simulating multiple iterations of induced-pluripotent stem cell cardiomyocyte models, Gong & Sobie [8] built populations of cell models and successfully translated their interactions with drugs onto human cardiomyocyte models through application of a Partial Least Squares Regression Analysis [9]. This same approach can be generalized to models of the same species but arising from different lineages, or translating across both species and lineages. The only requirement is that all models have a set of common (and corresponding) parameters that can be perturbed identically to simulate the same population-level variation.

Here we have applied these approaches to systematically understand the impact of mixing models from different lineages on translation performance. In particular, we sought to understand whether differences across model lineages imposed similar challenges to translation as the intrinsic electrophysiological differences across species. We found this to not be the case, and report that, at least for two distinct model lineages, species-translation works comparatively well across lineages as within. Importantly, this general finding did not hold for translation of action potential duration (APD) measures, and this may be due to different sensitivities of the model lineages to variations in the maximal transport rate of the Na^+-Ca^{2+} exchanger and Na^+-K^+-ATPase.

5.2 Methods

5.2.1 Models of Cardiac Electrophysiology

We modeled the action potential of a single ventricular cardiomyocyte for human, canine, and guinea pig species using the Washington University [3, 4, 5] (WU) and Simula [6] lineages. These models describe currents through voltage-gated ion channels in the form $I = go(v - E)$, where g is the channel conductance, v the membrane potential, E the channel's equilibrium potential, and o the channel's open probability which can be a function of either membrane potential or time-dependent gating variables. Figure 5.1 illustrates the currents, fluxes, and compartments available to each model and Table 5.1 lists which of these each model represents. All Simula models share the same formalisms and only differ in exact parameter values, while the WU models, though sequentially influenced by one another, were individually derived and feature notable differences in the ionic currents and cellular compartments they represent. Major similarities and differences include:

- All models split the sarcoplasmic reticulum (SR) into junctional and network compartments and the intracellular space into bulk cytosol and dyadic subspace. The two exceptions are that Simula models feature a third cytosolic compartment,

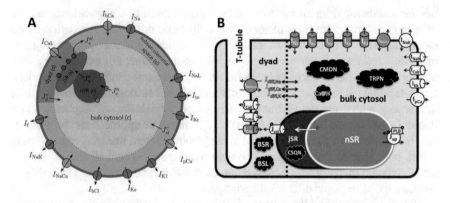

Fig. 5.1: Membrane currents (I), Ca^{2+} fluxes (J), and intracellular compartments for the (A) Simula model lineage, from [6], and (B) WU model lineage, from [5]. All WU models include the Ca^{2+} buffers calmodulin (CMDN) and troponin (TRPN) while only the WU human and canine models include calsequestrin (CSQN), anionic SR and sarcolemmal Ca^{2+} binding sites (BSR, BSL), and CaMKII which influences the gating of ion channels marked with a black ring in panel B. Both images in this figure are reproduced with permissions dictated by the Creative Commons Attribution (CC BY) license (https://creativecommons.org/licenses/by/4.0/).

the subsarcolemma (SSL), and the WU Guinea Pig model features only one cytosolic compartment.

- Simula models track free and buffered Ca^{2+} concentration ($[Ca^{2+}]$) while WU models only track free $[Ca^{2+}]$ but also model concentrations of compartment-specific Ca^{2+} buffers such as troponin and calmodulin.
- Simula models feature invariant [Na+], [K+], and [Cl-], while WU models treat these as dependent variables.
- WU human and canine models feature calcium/calmodulin-dependent protein kinase II (CaMKII) phosphorylation of target ion channels, as well as dual-component I_{NaCa}, where 80% operates in the bulk cytosol while the rest is dyadic.
- Simula models feature a Markovian representation of I_{Ks} while WU models do the same for I_{Kr}.

Baseline parameter and initial condition values were taken from each model's original publication. We generated populations of 1000 different parameter configurations for each model by multiplying the baseline value of a selection of common parameters (g_{K1}, g_{Kr}, g_{Ks}, g_{CaL}, g_{bCa}, g_{Na}, g_{NaK}, g_{NaCa}, g_{SERCA}, g_{RyR}) by a scaling factor sampled from the Guassian log-normal distribution with mean of 1 and 30% standard deviation; the same scaling factors were used for each model.

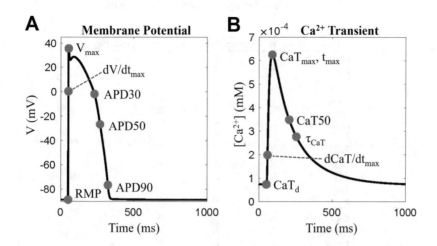

Fig. 5.2: (A) Membrane potential and (B) CaT traces depicting model features of resting membrane potential (RMP), maximum upstroke velocity (dV/dt$_{max}$), voltage amplitude (V$_{max}$), action potential duration to 30, 50, and 90% repolarization (APD30, 50, 90), diastolic $[Ca^{2+}]$ (CaTd), maximum CaT velocity (dCaT/dt$_{max}$), time and value of maximum CaT (CaT$_{max}$, t$_{max}$), time to 50% CaT decay (CaT$_{50}$), and CaT time constant (τ CaT).

5.2.2 Feature Extraction

Descriptors or features of each model's resulting electrophysiology were measured during steady state activity by applying a stimulus of -80mV for 20ms to the model 500 times at a frequency of 1Hz. Features were measured as their average over the final 10 stimulations, and included maximum upstroke velocity, resting membrane potential (RMP), voltage amplitude, Ca^{2+} transient (CaT) amplitude, maximum velocity, time constant, time to peak, and time to 50% decay, diastolic $[Ca^{2+}]$, and APD at 30, 50, and 90% repolarization (Figure 5.2). Model configurations were omitted from subsequent analysis if a convergent solution to their system of equations could not be reached or if their steady state behavior featured a voltage amplitude less than 5mV, resting membrane potential greater than -20mV, APD90 standard deviation greater than 10%, or change in CaT amplitude of more than 2% for total SR $[Ca^{2+}]$, cytosolic $[Ca^{2+}]$, or cytosolic $[Na^+]$.

Table 5.1: Currents represented in each model.

Current	Simula	WU Human	WU Canine	WU Guinea Pig
I_{CaL}	X	X	X	X
I_{bCa}	X	X	X	X
I_{nCa}				
I_{NaL}	X		X	X
I_{hNa}		X		X
I_{to1}	X	X	X	
I_{to2}			X	
I_{Kr}	X	X	X	X
I_{K1}	X	X	X	X
I_{Ks}	X	X	X	
I_{hK}		X		
I_{pK}			X	X
I_{hCl}	X	X	X	X
I_{NaCa}	X	X	X	X
I_{NaK}	X	X	X	X
I_{KCl}			X	
I_{NaCl}			X	
I_f	X			
I_{nsCa}				X

5.2.3 Sensitivity Analysis and Translation

Sensitivity analysis is a quantification of the correlation between input model parameters and resulting electrophysiological features and thus determines the functional dependence of model features on specific ionic conductances. Translation correlates features between pairs of models and thus translates the electrophysiological response of different models to identical parameter perturbations. Projection to latent structures, also referred to as partial least squares (PLS), is a form of matrix decomposition used for sensitivity analysis and translations of the models. The nonlinear iterative partial least squares algorithm [9] was used to calculate PLS regression coefficients, referred to as the B matrix from the matrix decomposition $Y = XB$. The decomposition for sensitivity analysis in matrix format is $Y_{Features} = X_{Parameterizations}B$, and for translations is $Y_{Features} = X_{Features}B$. For sensitivity analysis, the set of configurations for a model and the corresponding set of features are used as the X and Y matrices resulting in a B matrix for each of the six models in the two lineages. For translations, there are fifteen B matrices, one for each pair of models.

The B matrix represents the linear relationships between X and Y, meaning we cannot determine features from parameters or translate without loss of information quantified by the R2 values computed on the difference between the actual data values and the values predicted from multiplying a data set with its applicable B matrix. In practice, the B matrix can be utilized on parameter configuration vectors outside of the population to predict and translate linearly approximated feature vectors without solving the lumped conductance ordinary differential cardiac electrophysiological model, simplifying the analysis of other datasets.

5.3 Results

5.3.1 Model Translation

Translation performance, as measured by the average R^2 of all feature translations for each model combination, was generally high with all average R^2 values greater than 0.75 and most close to 1 (Figure 5.3A-B); model lineage, species, or translation direction did not appear to significantly affect this metric. When averaging R^2s based on whether the translation was across or with lineages, inter-lineage translations only modestly underperformed compared to intra-lineage translations with respective R^2 values of 0.859 ± 0.038 and 0.793 ± 0.073 (Figure 5.3C). Focusing on a subset of the translations for all combinations of WU, Simula, human, and canine models, this general trend no longer holds (Figure 5.3D-G). Instead, all R^2 values are well above 0.8 except for measures of APD (<0.6) which was translated well for only the Simula canine-human translation (Figure 5.3G).

5.3.2 Translation Discrepancies

A likely source of this discrepancy may come from irreconcilable differences in AP morphology between human and canine models (Figure 5.4A). Regardless of lineage, canine model APs tend to feature the classic notch-dome shape due to transient repolarization from inactivation of depolarizing inward sodium currents and activation of the transient outward potassium and sodium-calcium exchanger currents [10]. The WU canine model had the largest notch, causing some configurations to reach 30% of repolarization immediately following peak voltage, thus resulting in biphasic distributions of APD (Figure 5.4B-D). With this behavior entirely absent from human models, linear translation between the two species was significantly impeded (Figure 5.4E-G).

Sensitivity analyses of this subset of models revealed that the sign of APD regression coefficients for g_{NaK} and g_{NaCa} were the same for all models (Figure 5.5A-F, J-L) except WU canine (Figure 5.5G-I). This difference indicates that the WU canine model will have an entirely opposite change to APD given the same perturbation to these conductances. Such a fundamental difference in model behavior may also provide an explanation for poor inter-species translation.

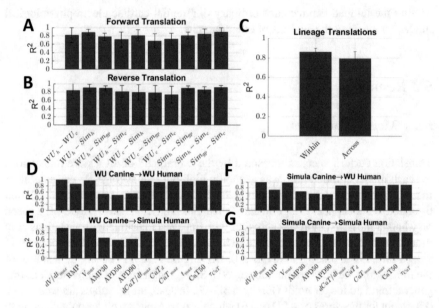

Fig. 5.3: Translation performance as measured by their resulting R^2 values. (A-B) Mean ± standard deviation R^2 values over all feature translations for each combination of human (h), canine (c), and guinea pig (gp) models. (C) Mean ± standard deviation R^2 values for translations within and across lineages for all species. (D-G) R^2 values for each feature of WU and Simula canine-human translation.

5.4 Discussion

Computational tools are increasingly seen as a key bridge for making the inferential jumps between electrophysiologic data collected among the major cardiac cell types used in drug screening (i.e. rodent, rabbit, canine, and human stem cell-derived cardiac myocytes). Because models of these cell types contain the basic properties of each cell type and differences between them, both linear and non-linear methods have been developed to translate among them, and importantly, to adult human cardiac cells. In this way, one holy grail of computational cardiac pharmacology is to leverage these models to reliably predict clinical outcomes of drugs based on data collected in other species or human cell lines.

Due to the inherent biological variability, it is important to use a technique that captures this variability. To this end, we developed a systematic approach of modifying the values of parameters common to all investigated models to obtain a population of 1000 configurations for each electrophysiological model. Resulting features of model behavior extracted from each variation of the model, such as APD and CaT velocity, were used to translate between models where the behavior of one model was predicted based on the measured behavior of another. Applying

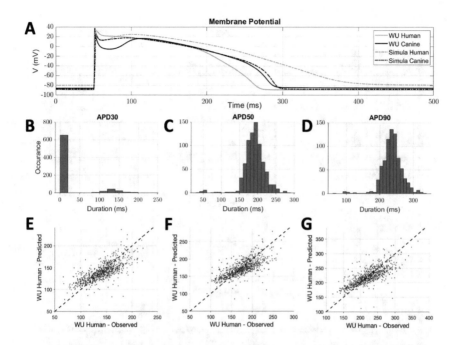

Fig. 5.4: (A) Steady state action potential traces generated using baseline parameter values for human (grey) and canine (black) models from both WU (solid) and Simula (dashed) lineages. (B-D) Histograms depicting the distributions of APD30 (29.1±48.7ms), 50 (191.0±29.8), and 90 (236.0±30.8) from the WU canine model measured as the average of the last 10 stimulations following 490s of 1Hz pacing. (E-G) Predicted APD30, 50, and 90 values for the WU human model using the regression coefficients from the WU canine – WU human translation analysis. Translation R^2 values for APD30, 50, and 90 are respectively 0.5268, 0.5014, and 0.5523. Black dashed lines are the identity line.

the NIPALS method for translation provided adequate predictive power as indicated by high R^2 values on average. Unexpectedly, there was not a large dependence of translation performance on model lineage, suggesting that their specific formulations might not be as large a source of inherent model behavior as initially thought. It appears that interspecies differences play a more significant role in affecting translation performance than the model's formulation.

As an illustrative example, we focused on canine to human translation between and within both lineages. All features were predicted well with the exception of ADP, except in the case of Simula canine to human translation which performed uniquely well. Sensitivity analyses revealed that the APD of WU canine models reacted oppositely to perturbations of conductances g_{NaK} and g_{NaCa} compared to all other models, indicating that blockers of the Na^+-Ca^{2+} or Na^+-K^+-ATPase exchanger

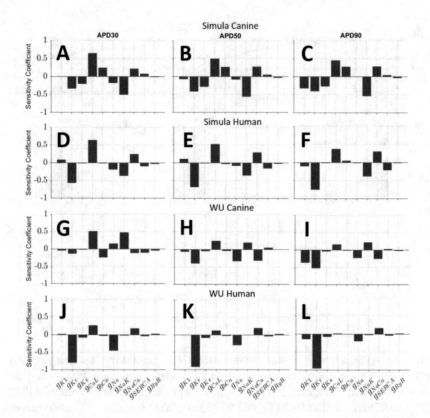

Fig. 5.5: Normalized sensitivity coefficients, grouped by APD30, 50 and 90 (columns), for the Simula canine (A-C), Simula human (D-F), WU canine (G-I), and WU human (J-L) models.

would have an opposite effect on the WU Canine model compared to the other models we investigated. Ultimately, the poor performance of APD translation implies that translation of electrophysiological behavior between species would inaccurately predict pathologies of APD such as QT interval prolongation and cardiotoxicity affecting the repolarization phase.

5.5 Conclusion

Here, we have proposed a ventricular cardiomyocyte model translation and parameter sensitivity analysis. We demonstrated that lineage dependence is not as strong as initially hypothesized, thus interspecies translation can be accurately performed between models of different origin. Most model features were robustly translated within and between model lineages, however, there were shortcomings in APD translation between human and canine models. Sensitivity analysis identified g_{NaK} and g_{NaCa} as potential candidates for the translation inconsistency where the WU canine model had an opposite APD reaction given the same perturbation to conductances of those transporters. This anomaly requires further investigation to ascertain whether it is due to that model's specific formulation or simply a general characteristic of the species.

References

1. Pablo Perel, Ian Roberts, Emily Sena, Philipa Wheble, Catherine Briscoe, Peter Sandercock, Malcolm Macleod, Luciano E Mignini, Pradeep Jayaram, and Khalid S Khan. Comparison of treatment effects between animal experiments and clinical trials: systematic review. *Bmj*, 334(7586):197, 2007.
2. HG Laverty, C Benson, EJ Cartwright, MJ Cross, C Garland, T Hammond, C Holloway, N McMahon, J Milligan, BK Park, et al. How can we improve our understanding of cardiovascular safety liabilities to develop safer medicines? *British journal of pharmacology*, 163(4):675–693, 2011.
3. Thomas J Hund and Yoram Rudy. Rate dependence and regulation of action potential and calcium transient in a canine cardiac ventricular cell model. *Circulation*, 110(20):3168–3174, 2004.
4. Ching-hsing Luo and Yoram Rudy. A dynamic model of the cardiac ventricular action potential. i. simulations of ionic currents and concentration changes. *Circulation research*, 74(6):1071–1096, 1994.
5. Thomas O'Hara, László Virág, András Varró, and Yoram Rudy. Simulation of the undiseased human cardiac ventricular action potential: model formulation and experimental validation. *PLoS computational biology*, 7(5):e1002061, 2011.
6. Karoline Horgmo Jæger, Verena Charwat, Bérénice Charrez, Henrik Finsberg, Mary M Maleckar, Samuel Wall, Kevin E Healy, and Aslak Tveito. Improved computational identification of drug response using optical measurements of human stem cell derived cardiomyocytes in microphysiological systems. *Frontiers in pharmacology*, 10:1648, 2020.
7. Aslak Tveito, Karoline Horgmo Jæger, Mary M Maleckar, Wayne R Giles, and Samuel Wall. Computational translation of drug effects from animal experiments to human ventricular myocytes. *Scientific Reports*, 10(1):1–11, 2020.
8. Jingqi QX Gong and Eric A Sobie. Population-based mechanistic modeling allows for quantitative predictions of drug responses across cell types. *NPJ systems biology and applications*, 4(1):1–11, 2018.
9. Paul Geladi and Bruce R Kowalski. Partial least-squares regression: a tutorial. *Analytica chimica acta*, 185:1–17, 1986.
10. Luis F Santana, Edward P Cheng, and W Jonathan Lederer. How does the shape of the cardiac action potential control calcium signaling and contraction in the heart? *Journal of molecular and cellular cardiology*, 49(6):901, 2010.

Open Access This chapter is licensed under the terms of the Creative Commons Attribution 4.0 International License (http://creativecommons.org/licenses/by/4.0/), which permits use, sharing, adaptation, distribution and reproduction in any medium or format, as long as you give appropriate credit to the original author(s) and the source, provide a link to the Creative Commons license and indicate if changes were made.

The images or other third party material in this chapter are included in the chapter's Creative Commons license, unless indicated otherwise in a credit line to the material. If material is not included in the chapter's Creative Commons license and your intended use is not permitted by statutory regulation or exceeds the permitted use, you will need to obtain permission directly from the copyright holder.

Chapter 6
A Computational Study of Flow Instabilities in Aneurysms

Nanna Berre[1], Gabriela Castro[2], Henrik Kjeldsberg[3], Rami Masri[4], Ingeborg Gjerde[3]

1 – Norwegian University of Science and Technology, Norway
2 – Dept. Mechanical Engineering, Pontifical Catholic University of Rio de Janeiro, PUC-Rio, Rio de Janeiro, RJ, Brazil
3 – Simula Research Laboratory, Norway
4 – Rice University, Texas, USA

Abstract The majority of hemorrhagic strokes are caused by cerebral aneurysms, harboured in a large portion of the human population. Currently, it is unclear what constitutes a dangerous aneurysm, and the risk of rupture of such aneurysms is challenging to quantify. Previous studies have shown that flow dynamics play an important role in the development of aneurysms. However, there is varying consensus on blood flow patterns and stability. In this work, we formulate a Reynolds–Orr method to quantify the stability of blood flow in arteries. Applying this method to blood flow in four different arterial models, we find that the blood flow therein is unstable under physiological conditions. We show the most unstable eigenmodes for each of these models and discuss how they potentially could help explain the initiation and growth of aneurysms.

Keywords: flow instability, aneurysm, perturbation, Reynolds–Orr Method

6.1 Introduction

Stroke is a leading cause of death, accounting for approximately 11% of all deaths worldwide in 2015 [1], and is currently the main cause of disabilities [2]. As reported in a large cohort study in Japan, 85% of hemorrhagic strokes are caused by cerebral aneurysms [3]. A cerebral aneurysm is an outward bulging of an artery, thought to form due to a localized weakness in the arterial wall. As an estimate, 5-8% of the population harbour an aneurysm [4, 5], with an annual risk of rupture at 1-2% [6].

The rupture of such aneurysms result in non-traumatic subarachnoid hemorrhage which often leads to death or disability [4].

Unruptured aneurysms are typically detected from neurovascular imaging obtained when diagnosing some unrelated issue. Due to the increasing availability of such imaging, clinicians are also detecting an increasing number of unruptured cerebral aneurysms [4]. This observation prompts difficult clinical decisions on whether one should monitor the aneurysm without treatment or proceed with surgical intervention [4]. This raises the question of what quantitative measures reliably determine rupture risk [7].

The dynamics of blood flow are thought to play a significant role in the initiation and propagation of lesions along an artery [8]. In particular, computational fluid dynamics (CFD) has been used to calculate the wall shear stresses (WSS) of the blood flow in an artery. Several CFD studies have successfully used abnormal WSS as a marker in retroactively classifying aneurysms according to their risk of rupture [9]. However, there has been observed significant differences between CFD results among different scientific groups [10]. In particular, the prediction of flow patterns and flow instability has been found to vary, which further impacts the computation of e.g. the WSS.

In simulations of blood flow in aneurysms, the flow itself is typically assumed laminar. However, experimental evidence has shown transitional and turbulent flow to occur in blood vessels such as the aorta [11] and carotid artery [12], and in saccular aneurysms in dogs [13] and humans [14, 15]. The presence of turbulent-like flow could significantly change the magnitude of the WSS.

In this work, we investigate flow instabilities in patient-specific geometries. More precisely, we use the Reynolds–Orr method to quantify the most unstable perturbation one can make to patient-specific aneurysm geometries. The Reynolds–Orr method, as revisited by Scott in [16], provides an exact relation between a baseflow and the kinetic energy of a perturbation. The analysis itself is nonlinear, with no linear approximations, and yields a well posed linear symmetric eigenvalue problem that can be solved via standard methods. Combining the open-source softwares FEniCS [17] and SLEPc [18], we find that the baseflow and corresponding eigenproblem can be implemented in less than 150 lines of code. The eigenvector solutions represent the most unstable perturbations one can make to the baseflow, while the eigenvalue indicates the growth rate of the resulting instability.

The main contributions of this report are as follows:

- The formulation of a Reynolds–Orr method for studying instability of flow in pipe-like domains.
- An open-source implementation [19] of this method based on FEniCS and SLEPc that is readily available for others to use.
- A comparison of the most unstable flow perturbations one can make in different types of domains with and without aneurysms (as illustrated in Fig. 6.1).

The rest of the paper is organized as follows: Section 6.2.1 introduces the Navier–Stokes equations with Dirichlet and traction boundary conditions. The variational formulation for this problem is also presented. The derivation of the kinetic energy relation is presented in Section 6.2.2 where the eigenvalue problem is recalled, followed by its discretization in Section 6.2.3. In Section 6.2.4, we present our methodology for solving the eigenvalue problem, followed by the results in Section 8.3. Finally, we present a discussion and potential future work in Section 8.4.

6.2 Methods

6.2.1 Baseflow equations

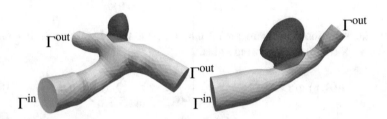

Fig. 6.1: Two arterial models, harbouring a terminal aneurysm (left) and a saccular aneurysm (right). The aneurysm is indicated in red.

Let $\Omega \subset \mathbb{R}^3$ denote an open, bounded flow domain with boundary $\partial\Omega$. Fig. 6.1 shows the two flow domains considered. The domain Ω is assumed to have an inflow boundary $\Gamma^{\text{in}} \subset \partial\Omega$ and an outflow boundary $\Gamma^{\text{out}} \subset \partial\Omega$. We denote by Γ the union of the inflow and outflow boundaries: $\Gamma = \Gamma^{\text{in}} \cup \Gamma^{\text{out}}$.

In Ω we then consider the steady incompressible Navier–Stokes equations for the fluid velocity $\boldsymbol{u}$ and pressure field p:

$$-\nu\Delta\boldsymbol{u} + \boldsymbol{u}\cdot\nabla\boldsymbol{u} + \nabla p = \boldsymbol{f} \quad \text{in} \quad \Omega, \tag{6.1a}$$

$$\nabla\cdot\boldsymbol{u} = 0 \quad \text{in} \quad \Omega, \tag{6.1b}$$

$$\boldsymbol{u} = \boldsymbol{0} \quad \text{on} \quad \partial\Omega\backslash\Gamma. \tag{6.1c}$$

In the equations above, ν is the kinematic viscosity, $\boldsymbol{f}$ is a given external force, and (6.1c) is a no-slip boundary condition to be applied on a subset of the boundary. We augment (6.1a) - (6.1c) by the following traction boundary conditions on Γ, namely

$$(\nu\nabla\boldsymbol{u} - pI)\boldsymbol{n} = p^i\boldsymbol{n}, \quad \text{on} \quad \Gamma^i \text{ for } i \in \{\text{in}, \text{out}\}, \tag{6.2}$$

where p^{in} and p^{out} are given data for the inflow and outflow boundaries, respectively.

Let $\boldsymbol{u}, \boldsymbol{v} \in H^1(\Omega)^3$. For the variational formulation, we first define

$$a(\boldsymbol{u},\boldsymbol{v}) = \nu \int_{\Omega} \nabla \boldsymbol{u} : \nabla \boldsymbol{v}, \tag{6.3}$$

$$c(\boldsymbol{u},\boldsymbol{u},\boldsymbol{v}) = \int_{\Omega} (\boldsymbol{u} \cdot \nabla \boldsymbol{u}) \cdot \boldsymbol{v}. \tag{6.4}$$

For $\boldsymbol{u} \in H^1(\Omega)^3$ and $p \in L^2(\Omega)$, we define

$$b(\boldsymbol{u},p) = \int_{\Omega} \nabla \cdot \boldsymbol{u} p. \tag{6.5}$$

We also define the following function space:

$$V_{\Gamma} = \{ \boldsymbol{v} \in H^1(\Omega)^d : \boldsymbol{v}|_{\partial\Omega\backslash\Gamma} = \boldsymbol{0} \}.$$

Then, the variational problem reads: find $(\boldsymbol{u},p) \in V_{\Gamma} \times L^2(\Omega)$ such that for all $(\boldsymbol{v},q) \in V_{\Gamma} \times L^2(\Omega)$ the following holds.

$$a(\boldsymbol{u},\boldsymbol{v}) + c(\boldsymbol{u},\boldsymbol{u},\boldsymbol{v}) - b(\boldsymbol{v},p) = \int_{\Omega} \boldsymbol{f} \cdot \boldsymbol{v} + \sum_{i \in \{\text{in,out}\}} \int_{\Gamma^i} p^i \boldsymbol{n} \cdot \boldsymbol{v}, \tag{6.6a}$$

$$b(\boldsymbol{u},q) = 0. \tag{6.6b}$$

6.2.2 Flow perturbations and instability

In this section, we will derive an eigenvalue problem that describes different perturbations one can make to this baseflow. This derivation is non-trivial with pressure boundary conditions. For this reason, we from now on consider Dirichlet boundary conditions for the flux for the entire boundary. In particular, let $(\boldsymbol{u},p)$ be the solution of the following time-dependent Navier–Stokes equations:

$$\partial_t \boldsymbol{u} - \nu\Delta\boldsymbol{u} + \boldsymbol{u} \cdot \nabla\boldsymbol{u} + \nabla p = \boldsymbol{f} \quad \text{in} \quad \Omega \times (0,T], \tag{6.7a}$$

$$\nabla \cdot \boldsymbol{u} = 0 \quad \text{in} \quad \Omega \times (0,T], \tag{6.7b}$$

$$\boldsymbol{u} = \boldsymbol{g} \quad \text{on} \quad \partial\Omega \times (0,T], \tag{6.7c}$$

$$\boldsymbol{u} = \boldsymbol{u}^0 \quad \text{on} \quad \Omega \times \{0\}. \tag{6.7d}$$

In the above, $\boldsymbol{u}^0$ is the solution of the baseflow equations (6.1a)-(6.1c). We derive the kinetic energy relation for the velocity $\boldsymbol{u}$ solving (6.7a)-(6.7d). The derivation presented here closely follows the derivation in [20]. Then, we state the eigenvalue problem which results from the energy relation. Let now $(\boldsymbol{w},q)$ solve the same equations (6.7a)-(6.7d) with the same boundary condition $\boldsymbol{g}$, but different initial data $\boldsymbol{w}^0$, such that $\boldsymbol{w}^0 \neq \boldsymbol{u}^0$. Define $(\boldsymbol{v},o)$ as the difference of the two solutions:

$$\boldsymbol{v} = \boldsymbol{u} - \boldsymbol{w}, \quad o = p - q. \tag{6.8}$$

Thus, (v, o) satisfy:

$$\partial_t v - v \Delta v + (u \cdot \nabla u - w \cdot \nabla w) + \nabla o = 0 \qquad \text{in} \quad \Omega \times (0, T], \qquad (6.9a)$$

$$\nabla \cdot v = 0 \qquad \text{in} \quad \Omega \times (0, T], \qquad (6.9b)$$

$$v = 0 \qquad \text{on} \quad \partial \Omega \times (0, T], \qquad (6.9c)$$

$$v^0 = u^0 - w^0 \quad \text{on} \quad \Omega \times \{0\}. \qquad (6.9d)$$

Following [20], we decompose the nonlinear terms as follows

$$
\begin{aligned}
u \cdot \nabla u - w \cdot \nabla w &= u \cdot \nabla u - u \cdot \nabla w + u \cdot \nabla w - w \cdot \nabla w \\
&= u \cdot \nabla v + v \cdot \nabla w = u \cdot \nabla v + v \cdot \nabla u - v \cdot \nabla v.
\end{aligned}
\qquad (6.10)
$$

Multiplying (6.9a) by v, using (6.10), and integrating over Ω yields

$$(\partial_t v, v) + a(v, v) + c(u, v, v) + c(v, u, v) - c(v, v, v) + b(v, o) = 0. \qquad (6.11)$$

Define the following function space:

$$V := \{v \in H^1(\Omega)^3 : \nabla \cdot v = 0 \quad \text{and} \quad v = 0 \quad \text{on} \quad \partial \Omega\}. \qquad (6.12)$$

It can be observed from Lemma 20.1 in [16] that for any $w \in V$ we have

$$c(w, v, v) = 0 \quad \forall v \in V.$$

In addition, since $\nabla \cdot v = 0$, we have

$$b(v, o) = \int_\Omega \nabla \cdot vo = 0.$$

Thus, we obtain the following

$$
\begin{aligned}
\frac{1}{2} \frac{\partial}{\partial t} \int_\Omega |v|^2 &= -v \int_\Omega |\nabla v|^2 - \int_\Omega (v \cdot \nabla u) \cdot v \\
&= -v \int_\Omega |\nabla v|^2 - \frac{1}{2} \int_\Omega v^t (\nabla u + \nabla u^t) v.
\end{aligned}
\qquad (6.13)
$$

From (6.13), we can make the following observation. The flow u is energy unstable at time $t = 0$ if there exists a $v_0 \in V$ such that

$$-\frac{1}{2} \int_\Omega v_0^t (\nabla u_0 + \nabla u_0^t) v_0 - v \int_\Omega |\nabla v_0|^2 > 0. \qquad (6.14)$$

To simplify, we define

$$\lambda_v = \frac{B_u(v, v)}{a(v, v)} \quad \text{with} \quad B_u(v, w) = \frac{1}{2} \int_\Omega v^t (\nabla u + \nabla u^t) w.$$

In the above definition, the form $a(v, v)$ is given in (6.3) and we recall that for $v \in V$:

$$a(v,v) = v \int_{\Omega} \nabla v : \nabla v = v \|\nabla v\|^2.$$

Using the above definitions and expressions, we obtain an equivalent instability condition to (6.14). Namely, the flow is unstable at time $t = 0$, if there is a $v_0 \in V$ such that

$$\lambda_{v_0} < -1.$$

In addition, we denote

$$\lambda = \inf_{0 \neq v \in V} \lambda_v. \tag{6.15}$$

It is clear that if $\lambda \geq -1$ then the flow is stable. On the other hand, if $\lambda < -1$, then the flow is unstable. In particular, from (6.13), there exists a v such that

$$\left(\frac{1}{\|\nabla v\|^2} \right) \frac{\partial}{\partial t} \int_{\Omega} |v|^2 = -2v(1 + \lambda_v) > 0.$$

We also recall Poincare's inequality which states that

$$\|v\|^2 \leq C_P \|\nabla v\|^2 \quad \forall v \in V.$$

Hence, we have the following relation:

$$\frac{\partial}{\partial t} \log \left(\int_{\Omega} |v|^2 \right) \geq -\frac{2v}{C_P} (1 + \lambda_v).$$

From the above, one can deduce that the most negative λ_v leads to the most unstable mode and this value is what we seek our computations.

It is shown in [20] that the solution of (6.15) solves the following eigenvalue problem. Find $(\lambda, v) \in (\mathbb{R}, V)$ such that

$$B_u(v,w) = \lambda a(v,w) \quad \forall w \in V. \tag{6.16}$$

For more details on the derivation of this eigenvalue problem, we refer to sections 5.2 and 5.3 in [20].

6.2.3 Discretization

We introduce the following discrete polynomial spaces to approximate solutions to the baseflow equations (6.1a)-(6.1c) and to the eigenvalue problem (6.16). Let V_h^k denote the space of C^0 piecewise polynomials of degree k on a regular mesh of the domain Ω. Define the vector valued polynomial space

$$V_{\Gamma,h}^k = \{v \in (V_h^k)^3 : v|_{\partial\Omega \backslash \Gamma} = 0\}.$$

We approximate the solutions to the variational form resulting from the baseflow equations: Find $(\boldsymbol{u}_h, p_h) \in (V_{\Gamma,h}^2, V_h^1)$ such that for all $(\boldsymbol{v}_h, q_h) \in (V_{\Gamma,h}^2, V_h^1)$:

$$a(\boldsymbol{u}_h, \boldsymbol{v}_h) + c(\boldsymbol{u}_h, \boldsymbol{u}_h, \boldsymbol{v}_h) - b(\boldsymbol{v}_h, p_h) = \int_\Omega \boldsymbol{f} \cdot \boldsymbol{v}_h + \sum_{i \in \{in, out\}} \int_{\Gamma^i} p^i \boldsymbol{n} \cdot \boldsymbol{v}_h, \quad (6.17a)$$

$$b(\boldsymbol{u}_h, q_h) = 0. \quad (6.17b)$$

To approximate the solutions to the eigenvalue problem (6.16), we define

$$V_h^k = \{\boldsymbol{v} \in (V_h^k)^3 : \boldsymbol{v}|_{\partial\Omega} = \boldsymbol{0}\}.$$

We find $(\lambda, \boldsymbol{v}) \in (\mathbb{R}, V_h^2)$ such that

$$B_{\boldsymbol{u}_h}(\boldsymbol{v}_h, \boldsymbol{w}_h) = \lambda a(\boldsymbol{v}_h, \boldsymbol{w}_h) \quad \forall \boldsymbol{w}_h \in V_h. \quad (6.18)$$

6.2.4 Computational Methodology

As discussed in the introduction, the aim of this work is to numerically investigate flow instabilities in aneurysms. We investigate this by studying the flow through four different arterial models.

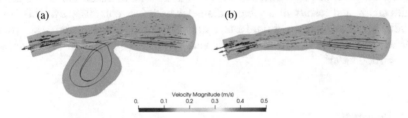

Fig. 6.2: Streamlines and vector field in the model with (left) and without (right) a saccular aneurysm, scaled and colored by the velocity magnitude.

To compute the most unstable flow modes for each of these models, we have followed the following procedure:

- **Step 1**: Mesh the geometry of a vessel with and without an aneurysm using Aneurysm Workflow developed by KVSlab [21].
- **Step 2**: Solve (6.17a) and (6.17b) for the baseflow $\boldsymbol{u}_h \in V_{\Gamma,h}^2$ in the meshed geometry using FEniCS [17].
- **Step 3**: Solve (6.18) for the eigenpairs $(\lambda, \boldsymbol{v}_h) \in (\mathbb{R}, V_h^2)$ with the computed flow $\boldsymbol{u}_h$ to determine the perturbations. This is done using SLEPc [18].

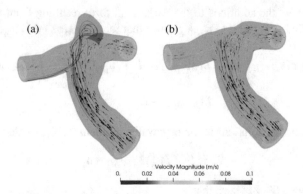

Fig. 6.3: Streamlines and vector field in the model with (left) and without (right) a terminal aneurysm, scaled and colored by the velocity magnitude.

The models have been collected from the Aneurisk database [22], and have been manually manipulated using Meshmixer [23]. To solve the baseflow equations (6.1a)-(6.2) and the eigenvalue problem (6.16), we used *FlowInstabilities* [19], an open-source CFD solver developed to investigate flow instabilities.

For the baseflow, we considered a laminar, steady state regime where the pressure drop was scaled so that the blood flow velocity is similar to the flow peak at systole. During the peak of the ventricular systole, the maximum distension of the artery wall occurs, and results in a greater diameter with less complacency. This lends validity to our assumption of rigid artery walls. The blood is considered to behave as a Newtonian fluid [24], with a kinematic viscosity of $v = 3.5 \cdot 10^{-6}$ m²/s and density of $\rho = 1060$ kg/m³.

6.3 Results

Figures 6.2 and 6.3 show the numerically computed baseflows. In addition to the vector field u, we also show the streamlines, i.e., the lines parallel to the velocity vector. Figure 6.2 shows the baseflow computed for the domains with and without a saccular aneurysm. The pressure drop across the domain was set to 0.001 mmHg, which resulted in a baseflow with peak magnitude of 50 cm/s. This corresponds well to values cited in the literature [25]. A small circulation is observed inside the aneurysm. This may indicate low values of WSS, which heightens the risk of disruption [26].

Figure 6.3 shows the baseflow computed for the domains with and without a terminal aneurysm. The pressure drop across the domain was set to 0.0001 mmHg, which resulted in a baseflow with peak magnitude of 10 cm/s. This is slightly lower than values cited in the literature. For increased pressure drops, however, the Newton solver did not converge. Examining the streamlines and the vector field of Fig. 6.3a and Fig. 6.3b, one sees a large recirculation in the area where the two different inlets meet. It is possible that for a higher pressure drop, no steady state solution to the Navier-Stokes equation exists for this branched geometry. For the model harbouring an aneurysm (Fig. 6.3a), the streamlines show a small flow circulation inside of it.

The most unstable eigenmodes for these baseflows are presented in Fig. 6.4 and Fig. 6.5. The figures also give the corresponding eigenvalue; an increase in absolute eigenvalue implies an increase in the kinetic energy of the instability. On the top row we present the most unstable mode, and on the bottom row the second most unstable mode, with its respective eigenvalue.

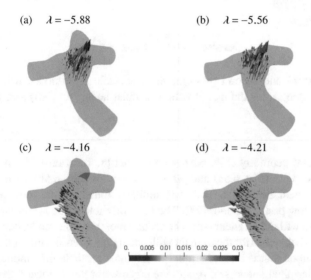

(a) $\lambda = -5.88$ (b) $\lambda = -5.56$

(c) $\lambda = -4.16$ (d) $\lambda = -4.21$

Fig. 6.4: Most (top) and second (bottom) most unstable eigenmodes with corresponding eigenvalue, λ, for the model with a terminal aneurysm (left) and without (right).

First considering the terminal aneurysm case in Fig. 6.4, we see that for both geometries (Fig. 6.4a and Fig. 6.4b) the most unstable eigenmode has an eigenvector directed to the point of the impingement of the jet located at the bifurcation. The corresponding eigenvalues are similar in magnitude. This is not unexpected as the baseflows for (6.4a) and (6.4b) are quite similar. Considering the impact of these perturbations on the baseflow, the most unstable perturbation points towards the aneurysm. Recalling that an aneurysm is caused by a weakening of the blood vessel wall, this may indicate that the flow perturbations and resulting instabilities play a role in aneurysm rupture. The eigenvectors of the second mode of instability for both cases (Fig. 6.4c and Fig. 6.4d) are similarly directed to the same location, near the entrance of the aneurysm.

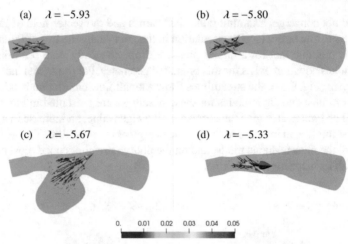

Fig. 6.5: Most (top) and second (bottom) most unstable eigenmodes with corresponding eigenvalue, λ, for the model with a saccular aneurysm (left) and without (right).

For the saccular geometry (Fig. 6.5), we find that the most unstable mode, for both the models - with (Fig. 6.5a) and without aneurysm (Fig. 6.5b) - are located downstream, i.e. near the outlet. This was similarly found for the most unstable flow modes for flow past a cylinder [27]. The most unstable mode is similar for the domains with and without an aneurysm. The second most unstable mode, conversely, shows a different behaviour for the domain with aneurysm. Considering Fig. 6.5c we observe a flow mode that is instead similar to the most unstable flow mode for the artery with the terminal aneurysm. Again, the eigenvector points toward the inside of the aneurysm. It is interesting to note that the second most unstable mode for the healthy geometry occurs in the constriction of the artery, see Fig. 6.5d. This could indicate that flow instabilities are the reason why the aneurysm occurs in this region.

6.4 Discussion

In this computational study, we used the theory of kinetic-energy instability [16] to analyze the most unstable flow perturbations one can make to blood flow in four different arterial models. The results indicate that blood flow in aneurysms can be unstable under physiological conditions. Moreover, the magnitude of the eigenvalue is seen to increase in the domains containing an aneurysm. This indicates that the kinetic energy of the instability increases if the domain contains an aneurysm. It is known, however, that the magnitude of the eigenvalue increases with the size of the domain, as discussed by [27]. Thus it is difficult to say if this effect is due to domain shape changes or simply occurs as the domain with aneurysm is larger.

For future work, it would be interesting to add the most unstable perturbations we computed to the baseflow and see the effect of this perturbation over time. In particular, it would be interesting to examine what effect the resulting perturbations has on the computation of the WSS.

References

1. Haidong Wang, Mohsen Naghavi, Christine Allen, Ryan M Barber, Zulfiqar A Bhutta, Austin Carter, Daniel C Casey, Fiona J Charlson, Alan Zian Chen, Matthew M Coates, et al. Global, regional, and national life expectancy, all-cause mortality, and cause-specific mortality for 249 causes of death, 1980–2015: a systematic analysis for the global burden of disease study 2015. *The lancet*, 388(10053):1459–1544, 2016.
2. Mira Katan and Andreas Luft. Global burden of stroke. In *Seminars in neurology*, volume 38, pages 208–211. Thieme Medical Publishers, 2018.
3. UCAS Japan Investigators. The natural course of unruptured cerebral aneurysms in a japanese cohort. *New England Journal of Medicine*, 366(26):2474–2482, 2012.
4. H Meng, VM Tutino, J Xiang, and A Siddiqui. High WSS or low WSs? complex interactions of hemodynamics with intracranial aneurysm initiation, growth, and rupture: toward a unifying hypothesis. *American Journal of Neuroradiology*, 35(7):1254–1262, 2014.
5. Gabriel JE Rinkel, Mamuka Djibuti, Ale Algra, and J Van Gijn. Prevalence and risk of rupture of intracranial aneurysms: a systematic review. *Stroke*, 29(1):251–256, 1998.
6. Gabriel JE Rinkel, Mamuka Djibuti, Ale Algra, and J van Gijn. Prevalence and risk of rupture of intracranial aneurysms. *Stroke*, 29(1):251–256, 1998.
7. Erdem Güresir, Hartmut Vatter, Patrick Schuss, Johannes Platz, Jürgen Konczalla, Richard Du Mesnil de Rochement, Joachim Berkefeld, and Volker Seifert. Natural history of small unruptured anterior circulation aneurysms: a prospective cohort study. *Stroke*, 44(11):3027–3031, 2013.
8. Alvaro Valencia, Alvaro Zarate, Marcelo Galvez, and Lautaro Badilla. Non-newtonian blood flow dynamics in a right internal carotid artery with a saccular aneurysm. *International Journal for Numerical Methods in Fluids*, 50(6):751–764, 2006.
9. Jianping Xiang, Sabareesh K Natarajan, Markus Tremmel, Ding Ma, J Mocco, L Nelson Hopkins, Adnan H Siddiqui, Elad I Levy, and Hui Meng. Hemodynamic–morphologic discriminants for intracranial aneurysm rupture. *Stroke*, 42(1):144–152, January 2011.
10. David A Steinman, Yiemeng Hoi, Paul Fahy, Liam Morris, Michael T Walsh, Nicolas Aristokleous, Andreas S Anayiotos, Yannis Papaharilaou, Amirhossein Arzani, Shawn C Shadden, Philipp Berg, Gábor Janiga, Joris Bols, Patrick Segers, Neil W Bressloff, Merih Cibis, Frank H Gijsen, Salvatore Cito, Jordi Pallarés, Leonard D Browne, and et.al. Variability of computational fluid dynamics solutions for pressure and flow in a giant aneurysm: The ASME 2012 summer bioengineering conference CFD challenge. *Journal of Biomechanical Engineering*, 135(2), February 2013.
11. Khalil Khanafer, Joseph Bull, Gilbert Upchurch, and Ramon Berguer. Turbulence significantly increases pressure and fluid shear stress in an aortic aneurysm model under resting and exercise flow conditions. *Annals of Vascular Surgery*, 21:67–74, 02 2007.
12. Seung Lee, Sang-Wook Lee, Paul Fischer, Hisham Bassiouny, and Francis Loth. Direct numerical simulation of transitional flow in a stenosed carotid bifurcation. *Journal of Biomechanics*, 41:2551–61, 08 2008.
13. Laligam Sekhar, Maomin Sun, D Bonaddio, and Robert Sclabassi. Acoustic recordings from experimental saccular aneurysms in dogs. *Stroke*, 21:1215–21, 09 1990.
14. Gary Ferguson. Turbulence in human intracranial saccular aneurysms. *Journal of Neurosurgery*, 33:485–97, 11 1970.
15. Yasushi Kurokawa, Seisho Abiko, and Kohsaku Watanabe. Noninvasive detection of intracranial vascular lesions by recording blood flow sounds. *Stroke*, 25:397–402, 03 1994.
16. L Ridgway Scott. *Introduction to Automated Modeling with FEniCS*. Computational Modeling Initiative LLC, 2018.
17. Martin Alnæs, Jan Blechta, Johan Hake, August Johansson, Benjamin Kehlet, Anders Logg, Chris Richardson, Johannes Ring, Marie E Rognes, and Garth N Wells. The FEniCS project version 1.5. *Archive of Numerical Software*, 3(100), 2015.
18. Vicente Hernandez, Jose E Roman, and Vicente Vidal. Slepc: A scalable and flexible toolkit for the solution of eigenvalue problems. *ACM Transactions on Mathematical Software (TOMS)*, 31(3):351–362, 2005.

19. Henrik Kjeldsberg, Rami Masri, Nanna Berre, Gabriela Castro, and Ingeborg Gjerde. Flowinstabilities. https://doi.org/10.5281/zenodo.5296829, August 2021.
20. L Ridgway Scott. Kinetic energy flow instability with application to Couette flow. Technical Report TR-2020-07, 2020.
21. Aslsk W Bergersen, Henrik A Kjeldsberg, and Kristian Valen-Sendstad. KVSlab. https://github.com/KVSlab, 2021.
22. Laura M Sangalli, Piercesare Secchi, and Simone Vantini. Aneurisk65: A dataset of three-dimensional cerebral vascular geometries. *Electronic Journal of Statistics*, 8(2):1879–1890, 2014.
23. Ryan Schmidt and Karan Singh. Meshmixer: an interface for rapid mesh composition. In *ACM SIGGRAPH 2010 Talks*, pages 1–1. 2010.
24. MO Khan, DA Steinman, and K Valen-Sendstad. Non-newtonian versus numerical rheology: Practical impact of shear-thinning on the prediction of stable and unstable flows in intracranial aneurysms. *International journal for numerical methods in biomedical engineering*, 33(7):e2836, 2017.
25. Benjamin S Aribisala, Zoe Morris, Elizabeth Eadie, Avril Thomas, Alan Gow, Maria C Valdés Hernández, Natalie A Royle, Mark E Bastin, John Starr, Ian J Deary, and Joanna M Wardlaw. Blood pressure, internal carotid artery flow parameters, and age-related white matter hyperintensities. *Hypertension*, 63(5):1011–1018, May 2014.
26. Y Zhang, L Jing, Y Zhang, and et al. Low wall shear stress is associated with the rupture of intracranial aneurysm with known rupture point: case report and literature review. *BMC Neurol*, 16(231), November 2016.
27. Ingeborg Gjerde and L Ridgway Scott. Kinetic-energy instability of flows with slip boundary conditions. *submitted*, 2021.

Open Access This chapter is licensed under the terms of the Creative Commons Attribution 4.0 International License (http://creativecommons.org/licenses/by/4.0/), which permits use, sharing, adaptation, distribution and reproduction in any medium or format, as long as you give appropriate credit to the original author(s) and the source, provide a link to the Creative Commons license and indicate if changes were made.

The images or other third party material in this chapter are included in the chapter's Creative Commons license, unless indicated otherwise in a credit line to the material. If material is not included in the chapter's Creative Commons license and your intended use is not permitted by statutory regulation or exceeds the permitted use, you will need to obtain permission directly from the copyright holder.

Chapter 7
Investigating the Multiscale Impact of Deoxyadenosine Triphosphate (dATP) on Pulmonary Arterial Hypertension (PAH) Induced Heart Failure

Kristen Garcia[1], Marcus Hock[1], Vikrant Jaltare[1], Can Uysalel[2], Kimberly J McCabe[3], Abigail Teitgen[1], Daniela Valdez-Jasso[1]

1 – Dept. of Bioengineering, University of California, San Diego, USA
2 – Dept. of Mechanical and Aerospace Engineering, University of California, San Diego, USA
3 – Simula Research Laboratory, Norway

Abstract 2-deoxy-ATP (dATP) is a myosin activator known to improve cardiac contractile force [1]. *In vitro* studies have shown that dATP alters the calcium transient profile in addition to the kinetics of the cross-bridge cycle [2]. Furthermore, *in vivo* studies of transgenic mice with increased production of dATP show elevated left ventricular systolic function [3]. Pulmonary arterial hypertension (PAH) is a rare disease of the pulmonary vasculature in which pressure overload in the right ventricle results in reduced contractile function and right heart failure [4]. We hypothesize that dATP may have a therapeutic effect on PAH-induced heart failure, by improving contractile function and restoring cardiac output and ejection fraction. However, because the effects of dATP cannot easily be assessed experimentally, we propose using a computational multiscale modeling approach to predict cardiac function. By altering parameters in an existing multiscale biventricular cardiac model [5], we were able to reproduce end-systolic and end-diastolic pressures and volumes that reflect the PAH condition, as well as healthy hearts. dATP was simulated by adjusting parameters in the model at the molecular and cellular levels based on experimental data [1], allowing us to predict the effects of dATP on PAH at the organ level. Our results show that the molecular effects of dATP can increase cardiac output and restore ejection fraction in PAH conditions, though at the cost of elevated mean arterial pressure, and may provide a new approach to treating this disease. Our multiscale modeling approach paves the way for further studies mapping out cardiovascular mechanics. As novel therapeutics continue to be discovered, their application and mechanism can be further explored through these computational models.

© The Author(s) 2022
K. J. McCabe (ed.), *Computational Physiology*, Simula SpringerBriefs on Computing 12, https://doi.org/10.1007/978-3-031-05164-7_7

7.1 Introduction

Pulmonary arterial hypertension (PAH) is a rare, yet aggressive disease with a 3-year survival rate of 54.9% for humans [6]. This life-threatening disease is characterized by pulmonary vascular remodeling, is represented by the narrowing of blood vessels in the lungs, and is indicated by a mean pulmonary arterial pressure (mPAP) above 20 mmHg [7]. PAH results in increased blood pressure which causes pressure overload in the right ventricle (RV) and leads to chamber remodeling [8]. The function and morphology of the RV has been found to have a very high correlation with the outcome of PAH. The RV initially adapts to the pressure overload by increasing muscle contractility and wall thickness in order to maintain cardiac output (CO), however, over time will maladapt and right heart failure (RHF) will occur leading to premature death. RHF is associated with increased filling pressures and dilation, along with reduced CO, ejection fraction (EF), and contractility [9]. Therapies such as vasodilator medications can alleviate the symptoms and improve quality of life for a short period of time [10], but do not work long term and eventually the RV will fail, leading to death.

2-deoxy-adenosine triphosphate, also known as dATP, is being studied as a potential treatment for heart failure [3]. dATP is a naturally occurring nucleotide that has been previously found to enhance crossbridge binding and cycling kinetics to improve contractility in muscle, when used to replace adenosine triphosphate (ATP) as the energy source [11]. Molecular modeling approaches have demonstrated that dATP induces allosteric changes in the myosin S1 fragment that lead to more favorable electrostatics and accelerated cycling kinetics [12, 13]. Furthermore, previous *in vivo* studies have demonstrated using a gene therapy approach to overexpress ribonucleotide reductase (RNR) [14] to restore function in animal models of heart failure, but focusing on the function of the left ventricle.

In this paper, we use a multi-scale modeling approach to determine if dATP can be used as a potential therapeutic for PAH. We combine cellular level models that include cross-bridge cycling and electrophysiology with an organ level lumped circulatory model and rat experimental data to predict cardiovascular function of the RV. We will discuss our *in silico* investigation of using dATP as a potential therapeutic for PAH with the use of multiscale models.

7.2 Methods

7.2.1 Cell Level Changes

Tewari *et al.* [15] have developed a cell level model of cardiomyocyte function described using a system of ordinary differential equations. While computationally relatively efficient, this model incorporates the essential components such as mitochondrial energetics, thin filament activation, and detailed description of cross-bridge kinetics. Furthermore, we have coupled this model to an electrophysiology model from Morotti *et al.* to produce an excitable model [16], while additionally incorporating a more detailed model of the SERCA pump [17]. As such, this model provides a means to replicate experiments via detailed simulations. In this work, we change specific model parameters to account for physical changes caused by the presence of dATP, as compared to ATP.

7.2.1.1 The SERCA Pump and Calcium transients

The Sarco/Endoplasmic Reticulum ATP-ase (SERCA) is an active pump that transports calcium-ions from the cytosol to the sarcoplasmic reticulum (SR). SERCA removes calcium-ions from the cytosol and contributes to maintaining the resting calcium-ion concentration in the 50 - 100 nM range [18]. In normal conditions, the SERCA pump consumes energy from the hydrolysis of 1 ATP molecule to transport 2 calcium-ions across the SR membrane [19, 17]. In addition, the SERCA also transports 2 H^+ out of the SR for every 2 Ca^{2+} flowing inside the SR [17, 20, 21].

Korte *et al.* [2] have shown that intracellular calcium transients undergo changes in the presence of dATP. The most notable feature of this dATP-induced effect is the shortening of the calcium decay time – defined as the time taken for the calcium-driven fluorescence indicator to reach half its peak value on stimulating the cardiomyocyte with a specific frequency – without having a significant effect on its peak amplitude. Korte *et al.* suggest that altered function of SERCA due to energetic effects of dATP on the pump rate could potentially explain this phenomenon but the mechanism still remains debated.

In the present study, we focused on the H^+/Ca^{2+} countertransport in SERCA to explain the increased Ca^{2+} pumping preferentially in the decay phase of the calcium-transient. Previous studies have reported changes in the H^+/Ca^{2+} countertransport due to conformational changes in the SERCA brought about by pH [22] and the presence of dATP [23]. To study the changes to H^+/Ca^{2+} countertransport, we varied the parameter n_H in the SERCA model from Tran et al. This parameter gives the stoichiometric coefficient of H^+ binding for state transitions $P_4 \rightarrow P_5$ and $P_8 \rightarrow P_9$ (see Figure 7.1). Based on the results from Tran *et al.*, we varied this parameter by $\pm 10\%$ to fit the experimental calcium transient from Korte *et al.*

7.2.1.2 Cross-bridge cycling kinetics

Tewari *et al.* [15] use a cross-bridge model composed of six states (super relaxed, non-permissive, permissive, weakly bound, strongly bound pre-powerstroke, and post-powerstroke). By altering the rate-constant parameters that describe transitions between states, the kinetics and force generation of the cross-bridge cycle can be modified to replicate the behavior observed in dATP conditions. In this work, we use experimental measurements from Force-pCa experiments and slack re-stretch (k_{tr}) of isolated rat trabeculae in the presence of ATP or dATP [1] to re-parameterize

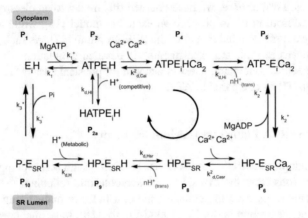

Fig. 7.1: SERCA pump reaction schematic. Figure adapted from Tran *et al.*, 2009 [17] with permission from Elsevier and Copyright Clearance Center (License number 5232480317559).

our cross-bridge model, in addition to using Brownian Dynamics (BD) simulation results from [13].

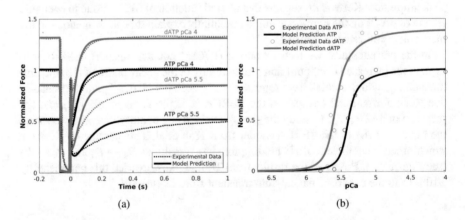

Fig. 7.2: (a) Force development measurements from experimental data and model simulations. ATP conditions shown in black, and dATP conditions in blue. Forces normalized to ATP experimental measurements at pCa 4. (b) Steady state normalized force development at different calcium concentrations in the presence of 5 mM ATP (black) and dATP (blue) from the model and experimental data.

Based on BD simulation results, the k_a association rate of the myosin head to actin was increased from 250 $\mu M s^{-1}$ to 567 $\mu M s^{-1}$. Similarly, the k_d dissociation rate was adjusted proportionately to remain thermodynamically constrained. In order to reduce overfitting, we manually optimized the fewest number of rate constants possible that could still characterize the change in behavior as seen by dATP for the force-pCa and slack re-stretch experiments. While this model does not reproduce exact measurements from the experimental data, relative changes between conditions provide an effective means to compare behavior. Figure 7.2 shows that at different calcium concentrations (pCa 4 and 5.5) dATP increases the rate of force redevelopment as compared to ATP. The exact rate measurements vary between the experimental data and model results, but the relative change in the rate constant is nearly proportional at both calcium concentrations (Table 7.1).

The model was also able to predict changes to the steady state force production in force-pCa experiments when in the presence of dATP compared to ATP. Most notably, the model predicted pCa_{50} and maximal force for dATP was significantly higher, matching the experimental results (Table 7.2). In total, only five new parameters were manually optimized to match dATP model behavior with experimental, and an additional three reverse rates adjusted to maintain proper cycle thermodynamics (Table 7.3).

Table 7.1: Slack Re-stretch k_{tr} Measurements.

	Experimental k_{tr}			Model k_{tr}		
Condition	ATP	dATP	**Relative Change**	ATP	dATP	**Relative Change**
pCa 4	13.4	16.04	**1.2**	20.1	24	**1.19**
pCa 5.5	2.9	5.9	**2.03**	9.37	15.7	**1.68**

Table 7.2: Force pCa Measurements.

	pCa 50		Max Normalized Force		Hill Coefficient	
	Experiment	Model	Experiment	Model	Experiment	Model
ATP	5.41*	5.39*	1*	1*	5.4	4.4
dATP	5.54*	5.56*	1.31*	1.4*	4.6	3.3

* p < 0.05 difference between ATP and dATP condition.

7.2.2 Organ Level Model

Previously developed models from Tewari *et al. (2016)*, Bazil *et al. (2016)*, Gao *et al. (2019)*, and a whole-organ cardiac mechanics model created by Lumens *et al. (2009)* are embedded in a lumped circulatory model that is used to simulate whole-body

Table 7.3: Cross-bridge Model Parameter Changes.

Condition	k_a	k_d*	k_1	k_2	k_3	K_{1SR}	k_{coop}	k_{-1}*	k_{-2}*
ATP	250	304.7	4	157	25	1.3	4	3	4
dATP	567	691.1	6	80	26	1.7	4.2	2	7.8

Note: denotes thermodynamically constrained.

cardiovascular function. TriSeg model created by Lumens *et al. (2009)* functions as a biventricular model which simulates left- and right- ventricular mechanics. Data from each rat is matched to the model which is run for 120 heart beats. Outputs of this model include values of mean arterial pressure and ejection fraction, along with plots for predicted right ventricular PV loops for the individual animal.

Experimental research around PAH is being conducted on Sprague Dawley rats in order to investigate right ventricular function throughout the progression of this disease. PAH is induced in rats following a Sugen-Hypoxia protocol, in which Sugen5416 (a vascular endothelial growth factor receptor (VEGFR) inhibitor) is injected into the rats following a 3-week period of hypoxia (10% oxygen). The rats are then removed from hypoxia and placed back into normoxia. The rats undergo open-chest terminal surgeries at different weeks throughout the disease progression. For the purpose of this project, we were focused on week 21 healthy and PAH induced male rats. During these open-chest surgeries, a pressure volume (PV) catheter is inserted into the apex of the RV to create *in vivo* PV loops. Several PV loops are recorded in steady-state conditions, with the average of these loops taken for the purpose of our analysis. The same process can be repeated while the PV catheter is in the left ventricle, however RV function and morphology is an important indicator for the severity of PAH making this ventricle the focus of this work.

Using the organ level model to simulate the pressure and flows in the whole-body [5], the goal was to replicate the experimental PV loops found during open-chest surgeries. The known values from the surgery day for blood pressure, heart rate, body weight, LV weight, RV weight, stroke volume, CO, LV diastolic volume, EF, and volume were inputted into the model and can be found in the supplementary information Table 0.8. We then found a number of adjustable parameters in the TriSeg (Heart) model that could be adjusted to replicate the experimental data PV loops. For both the healthy and PAH case, the stiffness of series element (K_{SE}), LV/RV/Septum wall volumes (Vw_{LV}, Vw_{RV}, Vw_{SEP}), and the LV/RV/Septal midwall reference surface areas ($Amref_{LV}$, $Amref_{RV}$, $Amref_{S}EP$) were adjusted to define the mass and geometry of the heart for each individual animal. Table 0.4 shows the relative changes of each of these parameters for the purpose of this work. The stiffness in the PAH model is 3-fold greater than that of the healthy case, representing a change that occurs in the RV during this disease. There are no original values for the $Amref_{LV}$, $Amref_{R}V$, $Amref_{S}EP$ parameters as these vary rat to rat.

We also adjusted inputs for each rat, such as the on rate constant for super relaxed state (K_{SR}), model parameter for force dependent super relaxed transition (K_{Force}),

Table 7.4: Adjusted Parameters for TriSeg (Heart) Model.

	Original Value	Healthy Rat Model	PAH Rat Model
K_{SE}	50000 mmHg/μm	80000 mmHg/μm	240000 mmHg/μm
Vw_{LV}	(LVW*2/3)/1000/1.05 mL	(LVW*2/3)/900/1.2 mL	(LVW*2/3)/1600/0.8 mL
Vw_{RV}	RVW/1000/1.05 mL	RVW/900/1.2 mL	RVW/1600/0.8 mL
Vw_{SEP}	(LVW/3)/1000/1.05 mL	(LVW/3)/900/1.2 mL	(LVW*2/3)/1600/0.8 mL
$Amref_{LV}$	N/A	2.1 cm^2	3.09 cm^2
$Amref_{RV}$	N/A	3.25 cm^2	1 cm^2
$Amref_{SEP}$	N/A	1.2 cm^2	4.1 cm^2

*LVW is the weight of the left ventricle and RVW is the weight of the right ventricle.

ATP hydrolysis rate (x_{ATPase}), and the resistances across PAH and healthy rats (R_{SA},R_{TAC}). These adjusted values, along with the inputs for each rat can be found in the supplementary information. Although, we were not able to replicate the loops exactly, we were able to get representative loops from the model that can be used to show relative changes from healthy to diseased PV loops. We were able to generate representative PV loops using the model that demonstrate relative changes between healthy and diseased conditions.

7.3 Results

We studied the impact of altered H^+/Ca^{2+} countertransport on intracellular calcium-transient by varying the parameter n_H from the Tran et al. [17] SERCA model. We found that an 8% increase in this parameter (from $2 \rightarrow 2.16$) was able to reproduce the shortening of decay-time observed in Korte et al. [2] without affecting the amplitude of the calcium-transient (Figure 7.3a). This change is consistent with the findings from Tran et al. which predicts $\pm 10\%$ [17] change to n_H during conformational changes to the SERCA pump. A caveat, however, is that we were able to observe a decrease in the decay-time of the calcium-transient by 33.4% as compared to about that of $\approx 50\%$ in Korte et al. [2] (see Figure 7.3b). This discrepancy could likely be due to difference in concentrations of metabolites like MgATP and MgADP between the experiment and the model and needs further investigation. Taken together, these results indicate that altered H^+/Ca^{2+} countertransport could potentially explain the results from Korte et al. [2] of the shortening of intracellular calcium transient decay-time without affecting the amplitude.

We combined the optimized parameters from the coupled SERCA and electro-hysiology with the altered sarcomere XB model to predict functional changes in cardiac muscle twitch. Because the XB cycling parameterization was carried out in 100% dATP conditions, while calcium handling measurements were carried out in approximately 2% dATP concentration, the optimized parameters are used in two different simulated conditions as seen in Figure 7.4. When the XB cycle parameters are used to simulate the 2% dATP case, they are scaled 2% linearly interpolated to the new fitted parameters in Table 7.3. The n_H parameter remains fixed at 2.16 in both cases. Comparable with experimental twitch observations, our results show that there is increased force production in the presence of dATP, as a result of altered XB kinetics. Furthermore, the twitch has an accelerated relaxation in the presence of dATP compared to ATP, due to adjustments in SERCA calcium handling parameters.

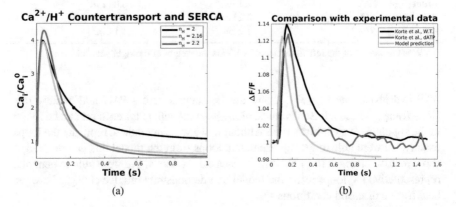

(a)   (b)

Fig. 7.3: (a) Calcium transient for 1 Hz stimulation for different values of n_H. (b) Comparison of calcium transient from model prediction with Korte *et al., 2011* data.

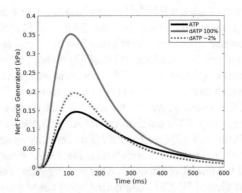

Fig. 7.4: Simulated cardiomyocyte twitch in the presence of ATP (black) and dATP (blue). The dotted line represents 2% simulated dATP and solid line represents 100% simulated dATP.

Mean arterial pressure (MAP) otherwise known as mPAP, is the determining metric for whether PAH has developed (MAP > 20 mmHg) or not (MAP < 20 mmHg). Other important factors when determining the function of the heart include EF, SV, and CO. Compared to the healthy case, in heart failure the MAP increases,

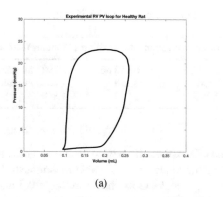
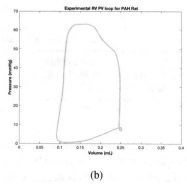

(a)   (b)

Fig. 7.5: Experimental rat right ventricle pressure volume loops for (a) Healthy and (b) PAH cases.

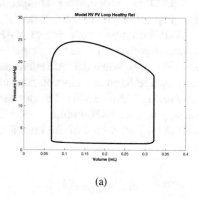
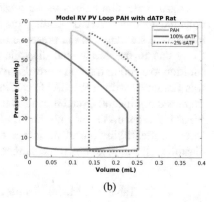

(a)   (b)

Fig. 7.6: (a) Model healthy rat right ventricle pressure volume loop. (b) Model PAH rat right ventricle pressure volume loops, without dATP treatment (yellow) and with dATP treatment (blue).

Table 7.5: Experimental versus Model Outputs for a Healthy Rat.

	Mean Arterial Pressure	Ejection Fraction	Stroke Volume	Cardiac Output
Model	11.555 mmHg	78.854 %	0.230 mL	75.757 mL/min
Experimental Data	17.940 mmHg	49.627 %	0.123 mL	38.742 mL/min
Ratio (model/exp)	0.644	1.589	1.870	1.990

Table 7.6: Experimental versus model outputs for a PAH Rat.

	Mean Arterial Pressure	Ejection Fraction	Stroke Volume	Cardiac Output
Model	35.962 mmHg	62.182 %	0.150 mL	35.987 mL/min
Experimental Data	39.730 mmHg	22.595 %	0.150 mL	19.009 mL/min
Ratio (model/exp)	0.905	2.752	1.000	1.890

accompanied by a decrease in EF, SV, and CO. Although we were unable to replicate the experimental PV loops perfectly using the model, we were able to demonstrate the increase in MAP and decrease in EF, SV, and CO as shown in Table 0.5 and Table 0.6. Parameter tuning around calculating the EF requires further investigation, as the values calculated were 1.5-fold greater for the model in the healthy case and almost 3-fold greater in the PAH case.

Organ scale simulations in the presence of dATP showed variable therapeutic potential to treat PAH depending on dosage (Figures 7.5b and 7.6). When PV loops were simulated in the presence of 2% dATP, there was a reduction of MAP, likely due to the accelerated diastolic relaxation, however, the cardiac output in addition to ejection fraction were also reduced (Table 7.7). When the PAH condition was simulated with 100% dATP, cardiac output was restored and ejection fraction improved significantly, yet this came at the cost of further elevated MAP. The elevated MAP indicates dATP may not be an ideal therapy to treat PAH. Future work will focus on identifying a critical dosage of dATP to take advantage of the improved diastolic relaxation and increase cardiac output.

Table 7.7: Model Results from dATP Organ Simulation.

	Healthy ATP	PAH ATP	PAH + 2% dATP	PAH + 100% dATP
MAP (mmHg)	11.6	36.0	26.4	48.1
EF (%)	78.9	62.2	45.1	93.4
SV (mL)	0.230	0.150	0.113	0.2116
CO (mL/min)	75.8	36.0	27.1	50.8

7.4 Discussion and Conclusion

In this paper, we have demonstrated that the cellular level models provide a framework to effectively simulate the behavior of dATP. Considering the interplay between our model and experimental data, there is a quantitative agreement of the theoretical results with the experimental observations, suggesting that our model captures the essential physics involved in the mechanics of PAH induced heart failure. However, there is a variation between experimental data and mathematical model. From these results, we can conclude the model replicates the experimental data with better precision for the PAH rat case compared with the healthy rat, although further parameter tuning is needed to advance the capabilities of the whole-organ model. Furthermore, the cell level dATP parameterization may require more detailed scaling approaches to accurately reflect how contraction changes as a function of dATP concentration. The next opportunity, in our view, lies in exploring this behaviour using the addition of noise terms by transforming our ODE model to Stochastic Differential Equation (SDE) model. A stochastic model will provide a method to study the beat to beat variability, and better characterize model stability.

From a modeling standpoint, series element resistance and cross-bridge kinetics play important roles in the mechanics of dATP on PAH induced heart failure. On the other hand, our model is not sensitive to variation in stiffness parameters and compliance parameters which are proximal aortic compliance, systemic arterial compliance, systemic venous compliance, and pulmonary arterial compliance. Because PAH is characterized by stiffening of the right ventricle, this is a significant limitation, and a more detailed model may be necessary to more accurately represent PAH. We showed that while this model can individually characterize the effects of PAH and dATP, it appears that dATP may have variable efficacy depending on the PAH conditions. As such, dATP may not be an ideal candidate to treat PAH, specifically given that dATP increases force and pressure in a high pressure disease. Additionally animal specific simulations are necessary for a robust prediction of dATP on function. This multiscale modeling approach shows that dATP is a promising therapeutic to treat PAH, however further detailed modeling and likely *in vivo* studies are necessary next steps.

7.5 Acknowledgements

The authors would like to acknowledge Prof. Andrew McCulloch, Dr. Kimberly McCabe, and Abby Teitgen for providing their critical comments and feedback for the project.

7.6 Supplementary Information

Table 7.8: Experimental Inputs for Rats on *Data1.xlsx*.

	Healthy Rat Data	PAH Rat Data
Body Weight (g)	640.9	646
LV Weight (mg)	484.11	593.73
RV Weight (mg)	340.6	776.6
HW (mg)	824.71	1370.33
LW (mg)	2341.38	3804.81
Heart rate (bpm)	329.38	239.91
SV	159.87	334.03
CO	52.66	80.14
LV diastolic Volume	500	408.02
EF (%)	74.47	81.87
V_{pre}	792.0006	853.5882
V_{post}	1052.722	24519.39
LV Vol d	416.7285	518.4661
LV Vol s	140.8056	284.4906
TAN	5.290799/7.623629	5.054202/7.282712
Crtot	21.00743/30.27007	18.36395/26.46103
Mito capacity	394.5743	342.2821

Table 7.9: *Adjustable_parameters_table_rest* Inputs.

	Healthy Rat Model	PAH Rat Model
$Amref_{LV}\ cm^2$	2.1	3.09
$Amref_{RV}\ cm^2$	3.25	1
$Amref_{SEP}\ cm^2$	1.2	4.1
K_{SR}	15.46914	21.511
K_{Force}	1.551729	2.1578
BV	1.101425	0.780733
$R_S A$	4	50
$R_T AC$	26	12.46694
x_{ATPase}	1.243375	1.074844

References

1. M Regnier, H Martin, RJ Barsotti, AJ Rivera, DA Martyn, and E Clemmens. Cross-bridge versus thin filament contributions to the level and rate of force development in cardiac muscle. *Biophysical Journal*, 87(3):1815–1824, September 2004.
2. F Steven Korte, Jin Dai, Kate Buckley, Erik R Feest, Nancy Adamek, Michael A Geeves, Charles E Murry, and Michael Regnier. Upregulation of Cardiomyocyte Ribonucleotide Reductase Increases Intracellular 2 deoxy-ATP, Contractility, and Relaxation. *Journal of molecular and cellular cardiology*, 51(6):894–901, December 2011.
3. Sarah G Nowakowski, Stephen C Kolwicz, Frederick Steven Korte, Zhaoxiong Luo, Jacqueline N Robinson-Hamm, Jennifer L Page, Frank Brozovich, Robert S Weiss, Rong Tian, Charles E Murry, and Michael Regnier. Transgenic overexpression of ribonucleotide reductase improves cardiac performance. *Proceedings of the National Academy of Sciences of the United States of America*, 110(15):6187–6192, April 2013.
4. Thenappan Thenappan, Mark L Ormiston, John J Ryan, and Stephen L Archer. Pulmonary arterial hypertension: pathogenesis and clinical management. *The BMJ*, 360:j5492, March 2018.
5. Rachel Lopez, Bahador Marzban, Xin Gao, Ellen Lauinger, Françoise Van den Bergh, Steven E Whitesall, Kimber Converso-Baran, Charles F Burant, Daniel E Michele, and Daniel A Beard. Impaired Myocardial Energetics Causes Mechanical Dysfunction in Decompensated Failing Hearts. *Function*, 1(2):zqaa018, September 2020.
6. Vallerie V McLaughlin, Sanjiv J Shah, Rogerio Souza, and Marc Humbert. Management of pulmonary arterial hypertension. *Journal of the American College of Cardiology*, 65(18):1976–1997, 2015.
7. Gérald Simonneau and Marius M Hoeper. The revised definition of pulmonary hypertension: exploring the impact on patient management. *European Heart Journal Supplements*, 21(Supplement_K):K4–K8, 2019.
8. Anton Vonk-Noordegraaf, François Haddad, Kelly M Chin, Paul R Forfia, Steven M Kawut, Joost Lumens, Robert Naeije, John Newman, Ronald J Oudiz, Steve Provencher, et al. Right heart adaptation to pulmonary arterial hypertension: physiology and pathobiology. *Journal of the American College of Cardiology*, 62(25S):D22–D33, 2013.
9. Cathelijne EE Van Der Bruggen, Ryan J Tedford, Martin Louis Handoko, Jolanda Van Der Velden, and Frances S de Man. Rv pressure overload: from hypertrophy to failure. *Cardiovascular research*, 113(12):1423–1432, 2017.
10. Stephen L Archer, E Kenneth Weir, and Martin R Wilkins. Basic science of pulmonary arterial hypertension for clinicians: new concepts and experimental therapies. *Circulation*, 121(18):2045–2066, 2010.
11. Joseph D Powers, Chen-Ching Yuan, Kimberly J McCabe, Jason D Murray, Matthew Carter Childers, Galina V Flint, Farid Moussavi-Harami, Saffie Mohran, Romi Castillo, Carla Zuzek, et al. Cardiac myosin activation with 2-deoxy-atp via increased electrostatic interactions with actin. *Proceedings of the National Academy of Sciences*, 116(23):11502–11507, 2019.
12. Sarah G Nowakowski, Michael Regnier, and Valerie Daggett. Molecular mechanisms underlying deoxy-adp. pi activation of pre-powerstroke myosin. *Protein Science*, 26(4):749–762, 2017.
13. Kimberly J McCabe, Yasser Aboelkassem, Abigail E Teitgen, Gary A Huber, J Andrew McCammon, Michael Regnier, and Andrew D McCulloch. Predicting the effects of datp on cardiac contraction using multiscale modeling of the sarcomere. *Archives of Biochemistry and Biophysics*, 695:108582, 2020.
14. Shin Kadota, John Carey, Hans Reinecke, James Leggett, Sam Teichman, Michael A Laflamme, Charles E Murry, Michael Regnier, and Gregory G Mahairas. Ribonucleotide reductase-mediated increase in datp improves cardiac performance via myosin activation in a large animal model of heart failure. *European journal of heart failure*, 17(8):772–781, 2015.
15. Shivendra G Tewari, Scott M Bugenhagen, Bradley M Palmer, and Daniel A Beard. Dynamics of cross-bridge cycling, ATP hydrolysis, force generation, and deformation in cardiac muscle. *Journal of Molecular and Cellular Cardiology*, 96:11–25, July 2016.

16. S Morotti, AG Edwards, AD McCulloch, DM Bers, and E Grandi. A novel computational model of mouse myocyte electrophysiology to assess the synergy between Na+ loading and CaMKII: CaMKII-Na+-Ca2+-CaMKII feedback in cardiac myocytes. *The Journal of Physiology*, 592(6):1181–1197, March 2014.

17. Kenneth Tran, Nicolas P Smith, Denis S Loiselle, and Edmund J Crampin. A Thermodynamic Model of the Cardiac Sarcoplasmic/Endoplasmic Ca2+ (SERCA) Pump. *Biophysical Journal*, 96(5):2029–2042, March 2009. Publisher: Elsevier.

18. Harjot K Saini and Naranjan S Dhalla. Modification of intracellular calcium concentration in cardiomyocytes by inhibition of sarcolemmal na+/h+ exchanger. *American Journal of Physiology-Heart and Circulatory Physiology*, 291(6):H2790–H2800, 2006.

19. Roberto Bravo, Valentina Parra, Damián Gatica, Andrea E Rodriguez, Natalia Torrealba, Felipe Paredes, Zhao V Wang, Antonio Zorzano, Joseph A Hill, Enrique Jaimovich, et al. Endoplasmic reticulum and the unfolded protein response: dynamics and metabolic integration. *International review of cell and molecular biology*, 301:215–290, 2013.

20. Daniel Levy, Michel Seigneuret, Aline Bluzat, and JL Rigaud. Evidence for proton countertransport by the sarcoplasmic reticulum ca2 (+)-atpase during calcium transport in reconstituted proteoliposomes with low ionic permeability. *Journal of Biological Chemistry*, 265(32):19524–19534, 1990.

21. Xiang Yu, S Carroll, JL Rigaud, and G Inesi. H+ countertransport and electrogenicity of the sarcoplasmic reticulum ca2+ pump in reconstituted proteoliposomes. *Biophysical journal*, 64(4):1232–1242, 1993.

22. Xiang Yu, Luning Hao, and Giuseppe Inesi. A pk change of acidic residues contributes to cation countertransport in the ca-atpase of sarcoplasmic reticulum. role of h+ in ca (2+)-atpase countertransport. *Journal of Biological Chemistry*, 269(24):16656–16661, 1994.

23. Kimberly J McCabe, Sophia P Hirakis, Abigail E Teitgen, Alexandre B Duclos, Michael Regnier, Rommie E Amaro, and Andrew D McCulloch. Exploring the effects of 2. deoxy-atp on serca 2a using multiscale modeling. *Biophysical Journal*, 118(3):260a, 2020.

Open Access This chapter is licensed under the terms of the Creative Commons Attribution 4.0 International License (http://creativecommons.org/licenses/by/4.0/), which permits use, sharing, adaptation, distribution and reproduction in any medium or format, as long as you give appropriate credit to the original author(s) and the source, provide a link to the Creative Commons license and indicate if changes were made.

The images or other third party material in this chapter are included in the chapter's Creative Commons license, unless indicated otherwise in a credit line to the material. If material is not included in the chapter's Creative Commons license and your intended use is not permitted by statutory regulation or exceeds the permitted use, you will need to obtain permission directly from the copyright holder.

Chapter 8
Identifying Ionic Channel Block in a Virtual Cardiomyocyte Population Using Machine Learning Classifiers

Bjørn-Jostein Singstad[1], Bendik Steinsvåg Dalen[2], Sandhya Sihra[3], Nickolas Forsch[4], Samuel Wall[4]

1 – ProCardio Center for Innovation, Dept. of Cardiology, Oslo University Hospital, Rikshospitalet, Oslo, Norway
2 – Dept. of Physics, University of Oslo, Norway
3 – Dept. of Bioengineering, University of California, San Diego, USA
4 – Simula Research Laboratory, Norway

Abstract

Immature cardiomyocytes, such as those obtained by stem cell differentiation, have been shown to be useful alternatives to mature cardiomyocytes, which are limited in availability and difficult to obtain, for evaluating the behaviour of drugs for treating arrhythmia. *In silico* models of induced pluripotent stem cell-derived cardiomyocytes (iPSC-CMs) can be used to simulate the behaviour of the transmembrane potential and cytosolic calcium under drug-treated conditions. Simulating the change in action potentials due to various ionic current blocks enables the approximation of drug behaviour. We used eight machine learning classification models to predict partial block of seven possible ion currents (I_{CaL}, I_{Kr}, I_{to}, I_{K1}, I_{Na}, I_{NaL} and I_{Ks}) in a simulated dataset containing nearly 4600 action potentials represented as a paired measure of transmembrane potential and cytosolic calcium. Each action potential was generated under 1 Hz pacing. The Convolutional Neural Network outperformed the other models with an average accuracy of predicting partial ionic current block of 93% in noise-free data and 72% accuracy with 3% added random noise. Our results show that I_{CaL} and I_{Kr} current block were classified with high accuracy with and without noise. The classification of I_{to}, I_{K1} and I_{Na} current block showed high accuracy at 0% noise, but showed a significant decrease in accuracy when noise was added. Finally, the accuracy of I_{NaL} and I_{Ks} classification were relatively lower than the other current blocks at 0% noise and also showed a significant drop in accuracy when noise was added. In conclusion, these machine learning methods may present a pathway for estimating drug response in adult phenotype

© The Author(s) 2022
K. J. McCabe (ed.), *Computational Physiology*, Simula SpringerBriefs on Computing 12, https://doi.org/10.1007/978-3-031-05164-7_8

cardiac systems, but the data must be sufficiently filtered to remove noise before being used with classifier algorithms.

8.1 Introduction

Drugs that act on the cardiovascular system may cause severe arrythmias, therefore animal or *in vitro* models are widely used for testing to validate the drug effect. The use of human induced pluripotent stem cell-derived cardiomyocytes (hiPSC-CMs) and microphysiological systems represents a new era in assessing drug-induced cardiotoxicity and screening of drug effects in microtissues [1, 2]. Even though hiPSC-CMs are one of the most developed and well-characterized model systems derived from induced pluripotent stem cells (iPSCs), there are still some limitations. Intraindividual variability, the lack of reproducibility of results across multiple laboratories, and a long maturation process are some of the biggest challenges in this field [2]. A recent study found the maturation process to take up to 8 months [3]. Another significant limitation is the small sample size that is being used in current studies [2], which limits the feasibility of using machine learning models. In this paper we show how simulated action potentials (AP) from *in silico* models can be used to train supervised machine learning models to classify partial ionic current blocks. Despite using simulated data sampled from a virtual population of hiPSC-CMs, this study demonstrates the feasibility of using machine learning models to classify ionic current block from action potential recordings.

8.2 Methods

This section describes the data used in this study and some pre-processing applied to the data. The different model architectures tested and the processes of model selection and hyperparameter tuning are described. Finally, the different scoring metrics used to analyze classifier performance are presented and explainable AI is discussed.

8.2.1 Data

The data set used in this study contained 4582 simulated action potentials from hiPSC-CMs under both normal conditions (control case, not drug-treated) and drug-treated conditions. Partial block of ionic currents was simulated in the cell model by reducing the maximal conductances of the corresponding channels. Each signal under drug-treated conditions was subtracted from the paired signal under normal conditions such that the resulting signal represented the change in the action potential due to drug treatment.

The seven current blocks that were simulated in this study were I_{CaL}, I_{Kr}, I_{to}, I_{K1}, I_{Na}, I_{NaL} and I_{Ks}. Figure 8.1 presents an overview of an action potential with different degrees of ion current block.

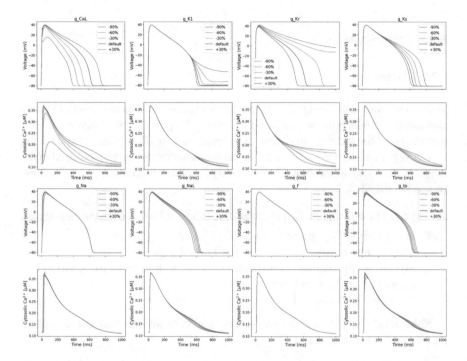

Fig. 8.1: Action potentials measured in terms of transmembrane voltage and cytosolic calcium with different percentages of ion current block. The f ion current block (shown in row three and four in the third column) is not present in the dataset used in this study.

Each simulated action potential was generated under $1Hz$ pacing and was represented as two signals: transmembrane voltage (V_t) and cytosolic calcium (Ca^{2+}). The two signals were represented as time-series with a length of 200, with 5ms between each time point.

The ground truth labels in the data set assumed that the ion channel was either blocked (1) or not blocked (0). Considering all seven ion currents, this gives a theoretical total of 128 unique combinations. All of the possible combinations were represented in the data set except for the case in which all currents are completely open, resulting in an actual total of 127 combinations.

$$2^7 - 1 = 127$$

The combinations were not equally represented in the data set; the least represented combination of ion current blocks had 27 examples, while the most represented combination had 40 examples. The mean and standard deviation were 36.1 and 2.1 respectively, which makes the data set slightly imbalanced and can translate to imbalanced classifier training.

8.2.2 Preprocessing

8.2.2.1 Noise

To provide more realistic signals as input to the classification models, random Gaussian noise was added to the simulated V_t and Ca^{2+} signals. Noise (within a certain percentage of the standard deviation of the mean across all signals in the data set) was incorporated into the raw signal by adding a vector of random noise of equal length as the signal to the signal itself. Listing 8.1 shows a code snippet of the noise generation function. [1]

Listing 8.1: The Python function that was used to add noise to the ion current signals.

```
def add_noise(X, percent =5.0):
    std = np.nanmean(X, axis =0).std()
    noise = np.random.normal(0, std ,X.shape)*percent/100
    X_noise = X + noise
    return X_noise
```

Noise levels of 0, 1, and 3 percent were added to all data and tested with all classifier models in this study.

8.2.2.2 Normalizing

After the addition of noise, the signals were normalized using min-max normalization. Equation 8.1 shows how x_i, which is the i-th element in X, can be normalized between 0 and 1 using the largest value in X, x_{max}, and the smallest value, x_{min}.

$$\tilde{x}_i = \frac{x_i - x_{min}}{x_{max} - x_{min}} \tag{8.1}$$

---

[1] The rest of the code developed in this study is available here: https://github.com/ SSCP2021-group-9/hiPSC-CMs-ionic-current-block-detecion

8.2.2.3 Subtract drug signals from control signals

After the addition of noise and the normalization of the signals between 0 and 1, the V_t signal from each ion's blocked case was subtracted from the V_t signal from the control case. The same was done for the Ca^{2+} signal.

8.2.2.4 V_t and Ca^{2+} concatenation

Finally the V_t and the Ca^{2+} signals were concatenated, meaning both V_t and Ca^{2+} for one simulated AP were appended into a single array. Figure 8.2 shows an example of the concatenation for all I_{Kr} current blocked signals with and without noise.

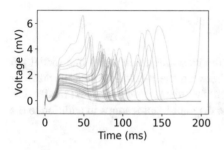

(a) Signal derived from transmembrane voltage with Kr block and 0% noise.

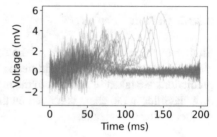

(b) Signal derived from transmembrane voltage with Kr block and 3% noise.

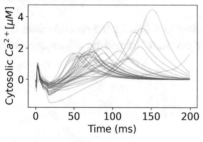

(c) Signal derived from cytosolic calcium with Kr block and 0% noise.

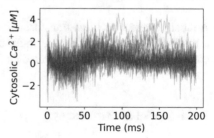

(d) Signal derived from cytosolic calcium with Kr block and 3% noise.

Fig. 8.2: Signals derived from the transmembrane voltage and cytosolic calcium signals after the preprocessing steps for the I_{Kr} channel. Figures (a) and (c) show the signals without noise, while (b) and (d) show the same signals with noise.

8.2.3 Multi-label classification methods

This section explains the three methods used to transform single-label classification methods into multi-label methods. These methods were implemented using Scikit-multilearn [4].

8.2.3.1 Binary relevance

The binary relevance method is considered to be the simplest strategy for problem transformation. It converts a multi-label problem into several independent binary classification problems [5].

8.2.3.2 Classifier chains

Similar to the binary relevance method, the classifier chain trains one model per class [6]. However, a notable difference between the methods is that all previous predictions are taken in addition to the input features to predict the next model. The n-th classifier in the chain is trained on the original input features in addition to the $n-1$ classification results.

8.2.3.3 Label Powerset

A label powerset problem transformation converts a multi-label classification into a multi-class problem with one multi-class classifier trained on all unique label combinations found in the training data [4]

8.2.4 Model architectures

8.2.4.1 Gaussian Naive Bayes

Naive Bayes (NB) is a series of classification methods that relies on Bayes theorem to make predictions, and Gaussian NB (GNB) is a variant of NB in which the data is assumed to follow a normal distribution.

If the input signal is $\mathbf{x}$, the likelihood of the signal belonging to a class C_k is

$$p(C_k \mid \mathbf{x}) = \frac{p(C_k)\, p(\mathbf{x} \mid C_k)}{p(\mathbf{x})} \propto p(C_k) \prod_{i=1}^{n} p(x_i \mid C_k). \qquad (8.2)$$

Since it would be the same for each class, $p(\mathbf{x})$ can be ignored. The proportion of training samples belonging to the class is $p(C_k)$. Further, the mean value μ_k and the standard deviation σ_k for a class C_k can be used in the formula for Gaussian probability distribution to calculate $p(x_i \mid C_k)$:

$$p(x_i \mid C_k) = \frac{1}{\sigma_k \sqrt{2\pi}} e^{-\frac{1}{2}\left(\frac{x_i - \mu_k}{\sigma_k}\right)^2} \tag{8.3}$$

The input is assumed to belong to the class with the highest likelihood [7].

This study used 3 versions of the GNB classifier: one version using the binary relevance method, another using the classifier chain method, and the last version using the label powerset method. The classifiers were implemented using sci-kit learn and scikit-multilearn [8, 4]. During training of these 3 GNB models the hyperparameters shown in Table 8.1 were tuned.

Parameter	Values
var smoothing	$[1, 10^{-1}, 10^{-2}, 10^{-3}, 10^{-4}, 10^{-5}, 10^{-6}, 10^{-7}, 10^{-8}]$

Table 8.1: Naive Bayes hyperparameters.

8.2.4.2 Support Vector Classifier

A Support Vector Machine (SVM) is a powerful method in machine learning. It is capable of doing linear and nonlinear classification, regression, and even outlier detection. A SVM used for classification purposes is sometimes called a Support Vector Classifier (SVC). SVCs are particularly well suited for classification of complex but small or medium-sized data sets. The mathematics behind the SVM/SVC relies on the definition of hyperplanes and the definition of a margin which separates classes of variables [9].

In this study the SVC was implemented in the label powerset algorithm using sci-kit learn. Table 8.2 shows the hyperparameters tuned in this study and the values used in the search space.

Parameter	Values
C	$[100, 1, 0.01]$
Kernel	[linear, poly, rbf, sigmoid]

Table 8.2: Support Vector Machine hyperparameters.

8.2.4.3 XGBoost

XGBoost is a scalable machine learning system for tree boosting [10]. This algorithm has shown promising results in many classification tasks. In this paper the XGBoost classifier is used inside the previously described label powerset algorithm. The model was implemented using a Python package developed by Chen, T. & Guestrin, C. 2016 [10].

Table 8.3 contains the tuned hyperparameters and the values used in the search space.

Parameter	Values
n estimators	$[1,5]$
min child weight	$[1,5]$
gamma	$[0.5,1]$
subsample	$[0.8]$
colsample bytree	$[0.8]$
max depth	$[3,5]$

Table 8.3: XGBoost hyperparameters.

8.2.4.4 Feed Forward Neural Network

Feed forward neural networks (FFNN), also called fully connected NN and multi-layer perceptrons, are the simplest form of neural networks. They consist of many nodes ordered into layers. The first layer is the input layer, which would contain the input features used in making a prediction. The last layer is the output layer, where each node represents the prediction for each of the labels. There are several layers between the input and output layers called hidden layers. Each node is connected to all the nodes in the previous layer, meaning that the value of a node depends on all the nodes in the previous layer. The strength of the connection depends on the internal weights and biases of the network, which are static for all predictions. The FFNN is trained using a process called backpropagation, which updates the internal variables using the gradient of the loss for a prediction [11].

An example of the structure of a FFNN can be seen in Figure 8.3.

The neural network was implemented in Python using Sci-kit learn [8]. Table 8.4 shows the hyperparameters that were tuned during the model development.

8.2.4.5 Convolutional Neural Network

The convolutional neural network used in this study was inspired by the the encoder model used in Fawaz HI et al 2019 [11]. The architecture of the encoder model is shown in Figure 8.4. The CNN model had 7 output neurons, equaling the number of

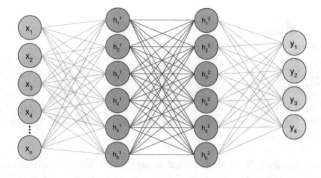

Fig. 8.3: An example of the structure of a feed forward neural network. Each node in one layer is connected to all the nodes in the next layer. Here we have *n* input features, 2 hidden layers with 6 nodes each, and 4 output nodes.

Parameter	Values
activation	ReLU
solver	[adam]
hidden_layer_sizes	(100,)
alpha	[0.0001]
batch size	[50, 100, 200]
learning rate init	[0.001]
learning rate	[adaptive]
max iter	[50, 100, 200]

Table 8.4: Neural Network hyperparameters.

classes in the dataset. A Sigmoid activation function was used in the final layer and binary cross-entropy was used as the loss function (Equation 8.4). The model was implemented in Python using Keras and Tensorflow [12, 13]. The hyperparameters tuned during training are shown in Table 8.5.

$$\mathcal{L}(y, \tilde{y}) = -(y \log \tilde{y} + (1 - y) log(1 - \tilde{y})) \tag{8.4}$$

Parameter	Values
epochs	[20, 60, 100]
batch size	[20, 60, 100]

Table 8.5: Convolutional Neural Network hyperparameters.

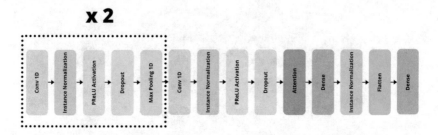

Fig. 8.4: The architecture of an encoder model. Each block illustrates a layer or a mathematical operation. The figure is obtained from B.Singstad et al. 2021 [14] with permission from the author.

8.2.4.6 Recurrent Neural Network

We developed a recurrent neural network (Figure 8.5) using a bidirectional long short-term memory (LSTM) block, a normal LSTM block, a dense layer with seven sigmoid outputs (one for each class), and a binary cross-entropy function as the loss function (Equation 8.4). The model was implemented in Python using Keras and Tensorflow [12, 13]. The hyperparameters that were tuned during training are shown in Table 8.6.

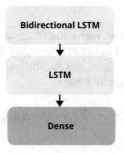

Fig. 8.5: The architecture of the recurrent neural network. Each block illustrates a layer or a mathematical operation.

Parameter	Values
epochs	$[20, 60, 100]$
batch size	$[20, 60, 100]$

Table 8.6: Recurrent Neural Network hyperparameters.

8.2.5 Model selection and hyperparameter tuning

The best models and hyperparameters were selected using nested cross-validation. A 10-fold stratified cross-validation was used as the outer loop and a regular 3-fold cross-validation was used as the inner loop. The number of inner loop cross-validations was equal to the number of possible combinations of hyperparameter settings in the search space. After being selected in the inner loop, the optimal hyperparameters were applied to the whole training data set in the outer loop.

8.2.6 Scoring and metrics

Accuracy, recall and precision were used to assess the performance of the eight models in this study.

8.2.6.1 Accuracy

The accuracy for each single channel (Equation 8.5) and the average accuracy across all channels (Equation 8.6) are both reported for each model. The accuracy score compares the predicted label ($\hat{y}$) with the true label (y). In Equations 8.5 and 8.6, n_s is the number of samples in the data set.

$$accuracy(y, \hat{y}) = \frac{1}{n_s} \sum_{i=0}^{n_s-1} 1 \cdot (\hat{y}_i = y_i) \tag{8.5}$$

In Equation 8.6, n_c is the number of classes from which the mean is calculated.

$$accuracy(y, \hat{y}) = \frac{1}{n_c} \sum_{j=0}^{n_c-1} \frac{1}{n_s} \sum_{i=0}^{n_s-1} 1 \cdot (\hat{y}_{c_i} = y_{c_i}) \tag{8.6}$$

8.2.6.2 Recall and precision

Precision and recall are two scoring metrics that give a number between 0 and 1. The precision score calculates the number of positive predictions that actually belong to the positive class, as seen in Equation 8.7. Recall, on the other hand, gives a score that quantifies the ratio of positive predictions made to all positive labels in the data set, as seen in Equation 8.8.

$$\text{Precision} = \frac{\text{TP}}{\text{TP} + \text{FP}} \tag{8.7}$$

$$\text{Recall} = \frac{\text{TP}}{\text{TP} + \text{FN}} \tag{8.8}$$

8.2.7 Explainable AI

Advanced AI models, such as CNN models, have been traditionally considered "black boxes" because until recently it has been impossible to explain their predictions [15]. However, a new discipline called explainable AI is making the decision basis more transparent. Explainable AI has the potential to reveal new features that could be of clinical importance. For example, explainable AI has been used to reveal new features on the electrocardiogram (ECG) that can potentially be used to detect particular gene mutations (Figure 8.6) [16]. We hypothesized that similar techniques would be helpful to highlight patterns in the data set used in this study, so local model agnostic explanations (LIME) were used to explain the predictions of the most advanced models.

8.2.7.1 LIME (Local Interpretable Model-Agnostic Explanations)

An example of a popular Python framework that can be used to implement local model agnostic explanations is LIME. By observing the classification result of a model, LIME makes small changes to the input data and trains a linear and explainable surrogate model. This process reveals the relative importance of the different properties in the input data [18].

In this study the linear surrogate model was trained to explain the CNN model. The surrogate model was trained on the same data set as the CNN model and validated on the same model as the CNN model.

8.3 Results

Figure 8.7 shows the results of the outer loop in the nested cross-validation, which was used to assess the accuracy of the models. The CNN model achieved the highest accuracy across all channels at three levels of noise with an average accuracy of 93% at 0% noise, 81% at 1% noise and 72% at 3% noise.

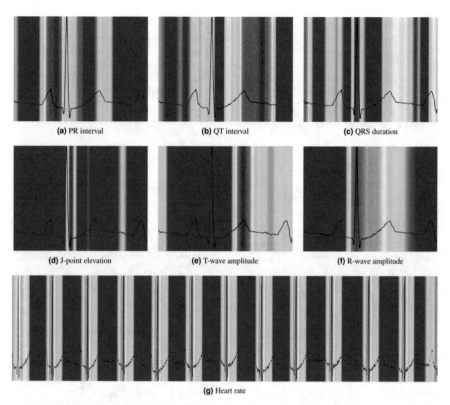

Fig. 8.6: Explainable AI techniques used on ECGs to explain features such as PR interval (a), QT-interval (b), QRS segment (c), J-point elevation (d), T-wave amplitude (e), R-wave amplitude (f) and heart rate (g). This figure is taken from [17] with permission from the authors and the publisher as determined by the Creative Commons CC BY license (https://creativecommons.org/licenses/).

8.4 Discussion

This study shows that signals derived from simulated action potentials can be used as inputs to train a supervised machine learning model and to classify partial block of specific ionic currents. The model that performed best in this study was a Convolutional Neural Network classifier, which achieved an average accuracy across all ion currents of 93% at 0% noise, 81% accuracy at 1% noise and 72% accuracy at 3% noise.

The findings shown in Figure 8.7 state that the CNN model had an accuracy close to 100% in classifying partial block of I_{CaL}, I_{K1}, I_{Kr}, I_{Na} and I_{to} when no noise was added. The I_{K1}, I_{Na} and I_{to} currents showed a significant drop in accuracy when noise was added. Conversely I_{CaL} and I_{Kr} showed little or no decrease in accuracy when noise was added. It seems that the classification of I_{CaL} and I_{Kr} is

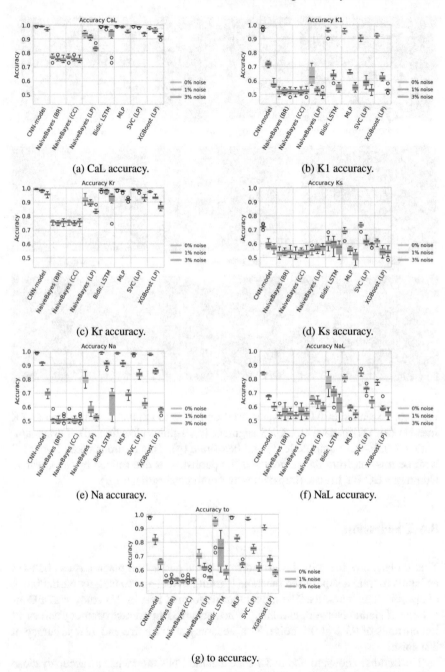

(a) CaL accuracy.

(b) K1 accuracy.

(c) Kr accuracy.

(d) Ks accuracy.

(e) Na accuracy.

(f) NaL accuracy.

(g) to accuracy.

Fig. 8.7: Classification accuracy for each individual ion channel represented as boxplots. The values shown represent the 10-fold cross-validated accuracy.

less vulnerable to noise than the I_{K1}, I_{Na} and I_{to} currents. It is possible that, as shown in Figure 8.1, I_{CaL} and I_{Kr} current block have a relatively large impact on the morphology of the action potential, while blocking I_{K1}, I_{Na} and I_{to} has less impact on the action potential. This observation may support the hypothesis that classification of ion current block is increasingly vulnerable to noise with decreasing impact of the current on action potential morphology.

The mean cross validated accuracy of I_{NaL} and I_{Ks} current block classification were 84% and 72% even at 0% noise. This was somewhat surprising because the changes in the action potential when I_{NaL} and I_{Ks} currents are blocked seem to be equally large or larger than action potentials with I_{Na} and I_{to} block in Figure 8.1. Experiments in this study show that the class activation for I_{Ks} is high for I_{CaL}, I_{Kr} and I_{to} blocks. The I_{NaL} block also shows high activation in signals labeled with I_{CaL}, I_{Kr} and I_{Ks}.

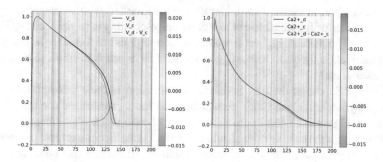

(a) I_{Ks} block interpreted as I_{Ks} block with a probability of 0.90.

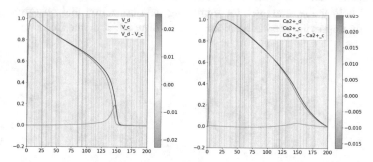

(b) I_{Ks} block interpreted as I_{Ks} block with a probability of 0.98.

Fig. 8.8: Activation map of the CNN classifier model for a time series input corresponding to sample cases with partial block of the I_{Ks} current only. The green line is the input to the classifier, representing the change in voltage and intracellular calcium due to partial block of I_{Ks}. Red regions along the time series indicate a positive activation for I_{Ks} block (label is 1), while blue regions indicate negative activation for I_{Ks} block (label is 0).

The high accuracy of the best classifier model, despite a large degree of variability within the virtual hiPSC-CM population, demonstrates the robustness of this machine learning approach. Figure 8.8 shows two sample cases of I_{Ks} partial block where the CNN classifier model correctly predicted the class with a probability of over 96%. Although the intracellular calcium signals have a different morphology, especially in the rate of repolarization, the classifier identified the correct channel block for both cases with a high degree of confidence. This is likely due to the similarities in the input time series, $V_{t_d} - V_{t_c}$ and $Ca^{2+}{}_d - Ca^{2+}{}_c$ (green line, Figure 8.8). Furthermore, the classifier was able to make a correct prediction based on an input with small values relative to the original voltage and intracellular calcium signals; the control and drug-treated traces were nearly indistinguishable to the human eye. This suggests the classifier may be robust for cases with both small and large degrees of channel block.

Figure 8.8 also shows the activation map of the CNN classifier model for both cases of I_{Ks} partial block, with red regions of the time series corresponding to positive activation for I_{Ks} block and blue regions corresponding to negative activation for I_{Ks} block. The darkest, highest frequency of red regions are found in the late phase of repolarization for both cases, suggesting that this region is the most important for determining class.

Figure 8.8a shows that the no-block classification is most pronounced for the voltage signal, while the block classification is most pronounced for the calcium signal. In Figure 8.8b the opposite is shown; the no-block classification is most pronounced for the calcium signal, while the block classification is most pronounced for the voltage signal. These conflicting results could suggest that both signals are relatively equal in importance when making a prediction.

The activation map provides meaningful insight into the input of the classifier and can help guide strategies for improving data collection and post-processing, ultimately improving prediction of the specific drug effect in question.

8.5 Conclusion

The purpose of this study was to investigate whether machine learning models are capable of detecting partial block of specific ionic currents from the change in morphology of an action potential. The results of this investigation show that a CNN model achieved the highest average accuracy in classifying ionic current block. The study also shows how the classifier performance is affected due to various levels of noise in the data. Although this study is based on simulated action potentials, the findings show the importance of filtering noise from measured signals, which may have implications in future studies that use action potential recordings of *in vitro* hiPSC-CMs. The study also showed that explainable AI methods can provide meaningful insight into the input of the classifier and can help guide strategies for improving both data collection and post-processing.

References

1. Anurag Mathur, Peter Loskill, Kaifeng Shao, Nathaniel Huebsch, SoonGweon Hong, Sivan G
    Marcus, Natalie Marks, Mohammad Mandegar, Bruce R Conklin, Luke P Lee, and Kevin E
    Healy. Human iPSC-based cardiac microphysiological system for drug screening applications.
    Scientific Reports, 5:8883, March 2015.
2. Li Pang. Toxicity testing in the era of induced pluripotent stem cells: A perspective regarding
    the use of patient-specific induced pluripotent stem cell–derived cardiomyocytes for cardiac
    safety evaluation. *Current Opinion in Toxicology*, 23-24:50–55, October 2020.
3. Yimu Zhao, Naimeh Rafatian, Nicole T Feric, Brian J Cox, Roozbeh Aschar-Sobbi, Erika Yan
    Wang, Praful Aggarwal, Boyang Zhang, Genevieve Conant, Kacey Ronaldson-Bouchard, Aric
    Pahnke, Stephanie Protze, Jee Hoon Lee, Locke Davenport Huyer, Danica Jekic, Anastasia
    Wickeler, Hani E Naguib, Gordon M Keller, Gordana Vunjak-Novakovic, Ulrich Broeckel,
    Peter H Backx, and Milica Radisic. A Platform for Generation of Chamber-Specific Cardiac
    Tissues and Disease Modeling. *Cell*, 176(4):913–927.e18, February 2019. Publisher: Elsevier.
4. Piotr Szymanski and Tomasz Kajdanowicz. Scikit-multilearn: a scikit-based Python environ-
    ment for performing multi-label classification. *The Journal of Machine Learning Research*,
    20(1):209–230, January 2019. arXiv:1702.01460 [cs.LG].
5. Gjorgji Madjarov, Dragi Kocev, Dejan Gjorgjevikj, and Sašo Džeroski. An extensive experi-
    mental comparison of methods for multi-label learning. *Pattern Recognition*, 45(9):3084–3104,
    September 2012.
6. Jesse Read, Bernhard Pfahringer, Geoff Holmes, and Eibe Frank. Classifier chains for multi-
    label classification. In *Joint European Conference on Machine Learning and Knowledge
    Discovery in Databases*, pages 254–269. Springer, 2009.
7. Marlis Ontivero-Ortega, Agustin Lage-Castellanos, Giancarlo Valente, Rainer Goebel, and
    Mitchell Valdes-Sosa. Fast Gaussian Naïve Bayes for searchlight classification analysis. *Neu-
    roImage*, 163:471–479, December 2017.
8. Fabian Pedregosa and et al. Scikit-Learn: Machine Learning in Python. *Journal of Machine
    Learning Research*, 12(85):2825–2830, 2011.
9. Durgesh K Srivastava and Lekha Bhambhu. DATA CLASSIFICATION USING SUPPORT
    VECTOR MACHINE. page 7, 2005.
10. Tianqi Chen and Carlos Guestrin. XGBoost: A Scalable Tree Boosting System. In *Proceedings
    of the 22nd ACM SIGKDD International Conference on Knowledge Discovery and Data
    Mining*, pages 785–794, San Francisco California USA, August 2016. ACM.
11. Hassan Ismail Fawaz and et al. Deep Learning for Time Series Classification: A Review. *Data
    Mining and Knowledge Discovery*, 33(4):917–963, July 2019.
12. Martín Abadi, Ashish Agarwal, Paul Barham, Eugene Brevdo, Zhifeng Chen, Craig Citro,
    Greg S. Corrado, Andy Davis, Jeffrey Dean, Matthieu Devin, Sanjay Ghemawat, Ian Good-
    fellow, Andrew Harp, Geoffrey Irving, Michael Isard, Yangqing Jia, Rafal Jozefowicz, Lukasz
    Kaiser, Manjunath Kudlur, Josh Levenberg, Dan Mane, Rajat Monga, Sherry Moore, Derek
    Murray, Chris Olah, Mike Schuster, Jonathon Shlens, Benoit Steiner, Ilya Sutskever, Kunal
    Talwar, Paul Tucker, Vincent Vanhoucke, Vijay Vasudevan, Fernanda Viegas, Oriol Vinyals,
    Pete Warden, Martin Wattenberg, Martin Wicke, Yuan Yu, and Xiaoqiang Zheng. TensorFlow:
    Large-Scale Machine Learning on Heterogeneous Distributed Systems. *arXiv:1603.04467
    [cs]*, March 2016. arXiv: 1603.04467.
13. Francois Chollet and et al. Keras, 2015. Publisher: GitHub.
14. Bjørn-Jostein Singstad. *Sammenligning av kardiologisk og algoritmebasert EKG-tolkning på
    idrettsutøvere: Kan kunstig intelligens forbedre algoritmene?* PhD thesis, February 2021.
15. Davide Castelvecchi. Can we open the black box of AI? *Nature News*, 538(7623):20, October
    2016. Section: News Feature.
16. RR van de Leur, K Taha, MN Bos, JF van der Heijden, D Gupta, MJ Cramer, RJ Hassink,
    P van der Harst, PA Doevendans, FW Asselbergs, and R van Es. Discovering and Visual-
    izing Disease-Specific Electrocardiogram Features Using Deep Learning: Proof-of-Concept

in Phospholamban Gene Mutation Carriers. *Circulation: Arrhythmia and Electrophysiology*, 14(2), February 2021. Number: 2 Place: United States.

17. Steven A Hicks and et al. Explaining deep neural networks for knowledge discovery in electrocardiogram analysis. *Scientific Reports*, 11(1):10949, May 2021. Bandiera_abtest: a Cc_license_type: cc_by Cg_type: Nature Research Journals Number: 1 Primary_atype: Research Publisher: Nature Publishing Group Subject_term: Cardiology;Machine learning Subject_term_id: cardiology;machine-learning.

18. Marco Tulio Ribeiro, Sameer Singh, and Carlos Guestrin. "Why Should I Trust You?": Explaining the Predictions of Any Classifier. *arXiv:1602.04938 [cs, stat]*, August 2016. arXiv: 1602.04938.

Open Access This chapter is licensed under the terms of the Creative Commons Attribution 4.0 International License (http://creativecommons.org/licenses/by/4.0/), which permits use, sharing, adaptation, distribution and reproduction in any medium or format, as long as you give appropriate credit to the original author(s) and the source, provide a link to the Creative Commons license and indicate if changes were made.

The images or other third party material in this chapter are included in the chapter's Creative Commons license, unless indicated otherwise in a credit line to the material. If material is not included in the chapter's Creative Commons license and your intended use is not permitted by statutory regulation or exceeds the permitted use, you will need to obtain permission directly from the copyright holder.

Printed in the United States
by Baker & Taylor Publisher Services